# Assessing Microbial Safety of Drinking Water

## IMPROVING APPROACHES AND METHODS

OECD

ORGANISATION FOR ECONOMIC CO-OPERATION AND DEVELOPMENT
WORLD HEALTH ORGANISATION

# ORGANISATION FOR ECONOMIC CO-OPERATION AND DEVELOPMENT

Pursuant to Article 1 of the Convention signed in Paris on 14th December 1960, and which came into force on 30th September 1961, the Organisation for Economic Co-operation and Development (OECD) shall promote policies designed:

- to achieve the highest sustainable economic growth and employment and a rising standard of living in member countries, while maintaining financial stability, and thus to contribute to the development of the world economy;
- to contribute to sound economic expansion in member as well as non-member countries in the process of economic development; and
- to contribute to the expansion of world trade on a multilateral, non-discriminatory basis in accordance with international obligations.

The original member countries of the OECD are Austria, Belgium, Canada, Denmark, France, Germany, Greece, Iceland, Ireland, Italy, Luxembourg, the Netherlands, Norway, Portugal, Spain, Sweden, Switzerland, Turkey, the United Kingdom and the United States. The following countries became members subsequently through accession at the dates indicated hereafter: Japan (28th April 1964), Finland (28th January 1969), Australia (7th June 1971), New Zealand (29th May 1973), Mexico (18th May 1994), the Czech Republic (21st December 1995), Hungary (7th May 1996), Poland (22nd November 1996), Korea (12th December 1996) and the Slovak Republic (14th December 2000). The Commission of the European Communities takes part in the work of the OECD (Article 13 of the OECD Convention).

# FOREWORD

Inadequate drinking water supply and quality and poor sanitation are among the world's major causes of preventable morbidity and mortality. According to the World Health Organization (WHO) estimates, basic hygiene-related diseases have a significant impact on human health. Diarrhoeal disease alone causes 2.2 million of the 3.4 million water-related deaths per year. Many of the deaths involve children under five years of age and the poorest households and communities. The problem is not limited to developing countries. In member countries of the Organisation for Economic Co-operation and Development (OECD), waterborne outbreaks occur all too frequently. Moreover, many outbreaks remain undetected, and it is likely that, beyond the reported outbreaks, there is an unrecognised background burden of disease.

Water-related issues were high on the international policy agenda in the 1970s, following the first international conference on the environment, held in Stockholm in 1972. However, by the time of the International Drinking Water Supply and Sanitation Decade (1981-90), interest had begun to wane. In the industrialised nations, concern focused on chemical contamination, and the international agenda moved increasingly towards major environmental issues such as global climate change, ozone depletion and desertification.

There was, however, an increasing level of public and professional concern about water safety, fuelled by concerns raised by outbreaks of disease and the recognition of new agents of disease and the challenges they presented to health protection. The 1993 Milwaukee outbreak, resulting in an estimated 400 000 cases of cryptosporidiosis, clearly underscored the severe consequences of waterborne outbreaks in OECD countries. The *Cryptosporidium* outbreak reported in Las Vegas, Nevada, in the spring of 1994 demonstrated the need for better understanding of the effectiveness of indicators and treatment processes in controlling waterborne pathogens. It also indicated the need for a re-evaluation of the effectiveness of traditional indicators as a basis for risk management, since the outbreaks occurred in waters that met the safety standards set by guidelines for traditional index and indicator bacteria.

3

Water and health have again moved up the international policy agenda as part of a more comprehensive understanding of sustainable development. This is evident in the declarations from the World Water Forums in Marrakesh (1997) and the Hague (2000) and in the increased co-operation among international agencies, including the programme of co-operation between the OECD and the WHO. The initiative leading to this report is a first product of that programme.

The need to improve assessment and management of the world's sources of drinking water was highlighted in 1996 at the OECD Workshop on Biotechnology for Water Use and Conservation in Cocoyoc, Mexico. Then, in 1998, the OECD Interlaken Workshop on Molecular Technologies for Safe Drinking Water reviewed the effectiveness of drinking water supply systems in protecting against microbial contaminants and the reliability of current monitoring parameters and testing systems. The Interlaken workshop confirmed the need for better microbial monitoring parameters and methods for assessing the safety of drinking water and monitoring and responding to adverse events. Most importantly, given the numbers of pathogens which cannot specifically be tracked by conventional methods, especially viruses and parasites such as *Cryptosporidium* and *Giardia*, the workshop recommendations pointed out that "business as usual" was no longer a viable option.

WHO's Guidelines for Drinking Water Quality provide a scientific basis for the development of standards and regulations to protect drinking water quality and human health. They are used by countries world-wide and are regularly updated in response to new information and developments. A series of meetings since 1995 has recommended adoption of a systematic preventive management approach to the control of microbial risks from catchment to consumer for drinking water. A framework integrating aspects of risk assessment and risk management in water safety was developed at a meeting in Stockholm (1999). The framework harmonises approaches applied to drinking water, wastewater use and recreational water quality. These include "Water Safety Plans", building upon Hazard Analysis Critical Control Point (HACCP) and the "multiple barrier principle". This document (developed by OECD and WHO) is one in a series of state-of-the-art reviews, which will inform the process of updating the Guidelines.

Outdated methods do not effectively identify and prevent serious enteric waterborne disease, and there is a large and under utilised toolbox for improving assessment of the safety of drinking water. While the rationale for using index organisms to detect contamination in source water remains sound, evaluation of treatment efficacy, post-treatment contamination, etc., require multiple indicators. No single microbial (or non-microbial) indicator parameter

4

is adequate to determine if all steps in the entire drinking water production process are working properly in all circumstances. Thus, it is necessary to gain a better understanding of the role and usefulness of the traditional and new parameters for monitoring and of the methods available for their analysis, and of the information needed to initiate appropriate remedial and preventive actions.

The Swiss Federal Institute for Environmental Science and Technology (EAWAG) heeded the call for a major review of the state of knowledge regarding monitoring parameters and testing methods relevant to the control of the microbial safety of drinking water. Under the leadership of the Director of the Institute, Professor Alexander Zehnder, and with the generous support of EAWAG, an initiative was launched to develop a guidance document to address such needs, in co-operation with the OECD and the WHO. Responsible for the co-ordination of this initiative were Dr. Mario Snozzi and Dr. Wolfgang Köster of EAWAG, Dr. Jamie Bartram of the WHO, Dr. Elettra Ronchi of the OECD and Dr. Al Dufour of the US Environmental Protection Agency. The successful outcome of this initiative is due, however, to the exceptional efforts made by the contributing international experts. The financial support of the Industry Council for Development (ICD) for review and advance document development is gratefully acknowledged. The expert editorial assistance of Dr. Lorna Fewtrell and the secretarial assistance of Alysia Ritter have been invaluable.

**Scope of the document**

This guidance document seeks to respond to the need to improve the assessment and management of the microbiological safety of drinking water, by moving away from using monitoring simply as a tool to verify the safety (or otherwise) of the finished product towards using the results as a basis for risk management actions. End-product testing comes too late to ensure safe drinking water, owing to the nature of current microbial sampling and testing, which typically provides results only after water has been distributed and often consumed. Thus, this document gives guidance on the appropriate application of monitoring parameters for ensuring the safety of drinking water and to inform risk management decisions, with an emphasis on control of faecal contamination. It offers guidance on how to select and use multiple parameters to meet specific information needs as a support to safe practice throughout the whole water system: catchment protection and assessment, source water quality assessment, assessment of treatment efficiency, monitoring the quality of drinking water leaving the treatment facility and in the distribution system. It offers a comprehensive review of traditional index and indicator organisms and of emerging technologies.

The approach described here has elements of both revolution and evolution. It is *revolutionary* in that it supports a rapidly emerging approach which emphasises the need to change from a single indicator organism, primarily used for end-product monitoring to determine hygienic quality, to multiple parameters including index and indicator organisms within a broader integrated management perspective and a risk management framework. It is *evolutionary* in that the approach builds upon the multiple barrier approach and on a body of information gleaned from scientific studies and surveys on the nature and behaviour of both pathogens and indicator organisms in water systems and on the relation between indicator organisms and pathogens.

Chapter 1 sets the scene, describing the problem and establishing the need for monitoring. It outlines the history of faecal indicator parameters and describes the various information needs. The use of a range of parameters to assess the examination of process efficiency and operational integrity is outlined in Chapter 2. Chapter 3 looks at the use of microbial monitoring in risk assessment. Chapters 4, 5 and 6 offer guidance on how the wide range of parameters can be put to use. Chapter 4 describes catchment characterisation and source water quality assessment, Chapter 5 looks at treatment efficiency and Chapter 6 examines the use of indicator parameters for monitoring the quality of drinking water during storage and distribution. Chapter 7 focuses on the investigation of water during incidents and disease outbreaks, with case studies illustrating the use of various parameters for specific purposes. Chapter 8 presents an overview of the various analytical techniques for determining numbers of faecal index and indicator bacteria as well as selected pathogens in water samples. It includes conventional and new (principally molecular biology) techniques and outlines the performance characteristics of the different methods along with their economic considerations (cost of infrastructure and consumables, level of training of technical staff).

**Challenges for the 21st century**

The document draws attention to important challenges related to the preservation and management of safe drinking water and particularly the need to develop a predictive system that will warn that a hazardous situation is imminent and enable timely and cost-effective correction of the situation. Perhaps the greatest challenge is the renewed recognition that resurgent and emerging pathogens with a high resistance to treatment are a significant hazard, not only in less developed countries, but in countries at all levels of industrialisation/economic development. Awareness of the existence of such organisms has developed primarily because of significant localised outbreaks. The specific aetiologic agent is only identified in about half of the detected outbreaks owing to the lack of appropriate detection methods or the lack of their application. Application of emerging molecular methods, while perhaps not appropriate for routine monitoring, are likely to make a significant contribution in this area.

The lack of available methodology to detect and quantify many such organisms, particularly those considered emerging waterborne pathogens, is an issue of great concern. Clearly, adequate assessment of the impact of such organisms on health is also directly related to the availability of appropriate detection methodology. In the post-genomics era, the tools for characterising microorganisms exist. Both genetic (nucleic-acid-based) and immunological tools are available and some molecular techniques appear particularly promising. For example, genotyping, or molecular characterisation, is a powerful new tool for identifying the source of microbial contaminants and is already in routine use for detecting *Cryptosporidium* in some countries. On the horizon, as Chapter 8 shows, are methods based on microarrays and biosensors. Advances in semiconductors and computers are expected to allow the next generation of microbial sensors to be small and simple devices, which are quick to respond. The future thus holds the promise of new techniques for detecting both existing and emerging pathogens.

Many challenges remain in the pursuit of safe drinking water for all. Resources are needed to increase the usefulness of the new molecular technologies in the pipeline. Advances in new molecular technologies should be encouraged and monitored, as they offer the best hope for improved and rapid detection of microbial contaminants in water.

# TABLE OF CONTENTS

*Chapter 1*

# SAFE DRINKING WATER: AN ONGOING CHALLENGE

*G.J. Medema, P. Payment, A. Dufour, W. Robertson,*
*M. Waite, P. Hunter, R. Kirby and Y. Andersson*

## 1.1     Introduction

### *1.1.1     Outbreaks of waterborne disease*

The microbiological quality of drinking water is a concern to consumers, water suppliers, regulators and public health authorities alike. The potential of drinking water to transport microbial pathogens to great numbers of people, causing subsequent illness, is well documented in countries at all levels of economic development. The outbreak of cryptosporidiosis in 1993 in Milwaukee, Wisconsin, in the United States provides a good example. It was estimated that about 400 000 individuals suffered from gastrointestinal symptoms due, in a large proportion of cases, to *Cryptosporidium,* a protozoan parasite (MacKenzie *et al.,* 1994), although subsequent reports suggest that this may be a significant overestimation (Hunter and Syed, 2001). More recent outbreaks have involved *E. coli* O157:H7, the most serious of which occurred in Walkerton, Ontario Canada in the spring of 2000 and resulted in six deaths and over 2 300 cases (Bruce-Grey-Owen Sound Health Unit, 2000). The number of outbreaks that has been reported throughout the world demonstrates that transmission of pathogens by drinking water remains a significant cause of illness. However, estimates of illness based solely on detected outbreaks is likely to underestimate the problem. A significant proportion of waterborne illness is likely to go undetected by the communicable disease surveillance and reporting systems. The symptoms of gastrointestinal illness (nausea, diarrhoea, vomiting, abdominal pain) are usually mild and generally only last a few days to a week, and only a small percentage of those affected will see a doctor.

11

Among these, only a minor proportion will have their stools microscopically examined and the examination usually starts with bacterial food pathogens. The number of reported outbreaks differs substantially among countries (Stenström, 1994) even from comparable ones, like the Scandinavian countries (Norway, Sweden, Denmark, Finland). In many cases, this is likely to reflect the effectiveness of the reporting systems rather than a true difference in the number (or size) of outbreaks.

Most sporadic cases of waterborne intestinal illness will not be detected or, if detected, may not be recognised as water-related. In industrialised countries, drinking water that meets current water quality standards may still harbour low concentrations of pathogenic microorganisms. These will cause occasional illness throughout the community served. It is very difficult to relate these sporadic cases to drinking water, as they are buried within the endemic level of disease circulating in the population through other routes of transmission (person-to-person, food and animal contact). There are, however, data from epidemiological studies and seroprevalence studies that indicate that endemic transmission of disease through drinking water does occur (Payment *et al.,* 1991, 1997; Isaac-Renton *et al.,* 1996).

### 1.1.2     The disease burden is high

Several researchers have attempted to estimate the total burden of waterborne disease world-wide. Huttly (1990) reported a total number of 1.4 billion annual episodes of diarrhoea in children under five years of age, with an estimated 4.9 million children dying as a result (although these were due to all causes of diarrhoea and not just water-related cases). While Prüss *et al.,*(2002) estimated that water, sanitation and hygiene was responsible for 4.0% of all deaths and 5.7% of the total disease burden occurring worldwide (accounting for diarrhoeal diseases, schistosomiasis, trachoma, ascariasis, trichuriasis and hookworm disease). Clearly, in countries where a large part of the population does not have access to safe drinking water, a substantial number of these infections will be waterborne, indeed, Hunter (1997) estimated that waterborne disease might account for one-third of the intestinal infections world-wide.

Waterborne disease is not restricted to developing countries. Morris and Levine (1995) attempted to estimate the annual waterborne disease burden in the United States of America and indicated that 560 000 people may suffer from a moderate to severe waterborne infection and that 7.1 million suffer from a mild to moderate waterborne infection each year. All waterborne infections may lead to an estimated 1 200 deaths a year. Even if these rough figures are overestimated, both the health and economic burden are considerable even for an industrialised society (Payment, 1997).

The diseases most frequently associated with water are enteric infections (such as infectious diarrhoea) that are also often associated with food (Mead *et al.*, 1999). In many cases, the disease is relatively mild and self-limiting. However, a proportion of the infected population will suffer more severe outcomes, especially when the health care system is lacking. Several waterborne pathogens, such as *Vibrio cholerae*, hepatitis E virus and *Escherichia coli* O157:H7, have high mortality rates (Hunter, 1997). In recent cholera outbreaks, for example, the fatality rate was generally 1-3%, but could be as high as 8-22%. Hepatitis E virus infections may lead to fulminant fatal hepatitis in 1-2% of cases, with pregnant women being especially at risk of more severe disease. Waterborne infections with *E. coli* O157:H7 are associated with haemorrhagic colitis and haemolytic uremic syndrome, both serious illnesses, with the latter occurring particularly in children. The fatality rate in waterborne outbreaks is 0.3-1.6% (Hunter, 1997; Bruce-Grey-Owen Sound Health Unit, 2000).

In the 1990s, evidence that microbial infections are associated with chronic disease started to accumulate. Several waterborne pathogens have been associated with serious sequellae (*i.e.* severe illness or chronic or recurrent disease that appears long after the initial exposure to contaminated water). Examples of sequellae that could potentially be associated with acute waterborne disease include:

- Diabetes, which has been linked to Coxsackie B4 virus (Roivainen *et al.*, 2000; Horwitz *et al.*, 1998).

- Myocarditis, which has been linked to echovirus (Ferreira Jr. *et al.*, 1995; Shanmugam *et al.*, 1986).

- Guillian-Barré syndrome associated with *Campylobacter* spp. (Prendergast and Moran, 2000).

- Gastric cancer, which has been linked to *Helicobacter* sp. (Uemura *et al.*, 2001).

- Reactive arthritis, which has been linked to *Klebsiella* sp. (Ebringer and Wilson, 2000).

With the exception of *Klebsiella*, the association of these microbes with acute waterborne disease has been well established. More remote connections between waterborne microbes and chronic disease has not been fully established, but is highly suspected (Hunter, 1997).

### 1.1.3    New pathogens emerge

Patterns of infection change over time and public health authorities can be been faced with newly discovered or emerging pathogens that may be able to overcome the many barriers of the water treatment and distribution systems. Emerging pathogens are defined as microorganisms which are responsible for infectious diseases and which have appeared or increased in occurrence during the past two decades (CDR, 1998). The issue of emerging pathogens came to the fore in the 1990s when water suppliers were shocked by the discovery of previously, essentially, unknown microorganisms, which were responsible for a series of waterborne outbreaks of illness. This is likely to continue into the future as such emergence or re-emergence has been linked to intensive agriculture, increased growth and migration of human populations and climate change (US Department of Health and Human Services, 1998; WHO, 1998). Examples of enteric waterborne emerging pathogens include caliciviruses, *E. coli* O157:H7, *Helicobacter* sp., *Mycobacterium avium* complex (MAC) and the protozoa *Cryptosporidium* sp., *Cyclospora* sp. and *Toxoplasma* sp. This problem requires constant vigilance in terms of what may pose a 'new threat' and also constant development with regard to methodologies and techniques for the detection of such threats. As noted by LeChevallier *et al.* (1999a), *"knowledge is the first line of defense toward providing safe drinking water."*

## 1.2    A history of making water safer

The recognition, in the 1800s that bacteria were agents of disease, along with the development of bacteriology as a science made it possible to use bacteria as tools to evaluate water quality and treatment. Essentially, non-pathogenic, easily detectable microorganisms were used to 'indicate' that contamination had taken place and, as such, there was a risk to public health. The need to be able to assess water quality, the fact that the majority of pathogens in drinking water are faecally derived and the 'moving' target presented by pathogens resulted in the idea to measure general levels of faecal contamination and the birth of the 'indicator' concept.

The presence of heterotrophic bacteria, measured by colony count following growth on a gelatine medium, has been used since the late 19th century to monitor general water quality as well as the function and efficiency of slow sand filtration. Koch (see Box 1.1) postulated that if the effluent of a slow sand filter contained less than 100 bacteria/ml, the water was suitable for drinking and presented no risk of cholera or typhoid. A number of findings paved the way for this development. The book *Microorganisms in Water*, for example, published by the Franklands in 1894 contained several important findings, including:

- The number of bacteria in water is a measure of pollution, and the number of bacteria in seawater, groundwater and lakewater should be below 100/ml.

- Slow sand filtration reduces the number of bacteria from river water by more than 90% to below 100/ml.

The 100 bacteria/ml level became a standard in many European countries where it was accepted as an attainable objective, while the USA and Canada adopted a 500 bacteria/ml guideline. Although the level of heterotrophic bacteria in drinking water is not related to contamination by pathogens, it is still present in most national laws on drinking water quality as an indicator of the overall quality of the water (van der Kooij, 1993; Anon, 1999).

---

**Box 1.1. Preventing disease transmission: the early years**

As early as 400 BC, it was recognised that polluted water was associated with disease (Whitlock, 1954). The first demonstration that disease was transmitted by water took place over 2000 years later, when the first cholera pandemics, which originated in India, struck Europe and resulted in many victims. At the time, it was generally believed that the disease was spread through bad odours. Preventive measures were taken against such odours.

John Snow, a prominent epidemiologist, studied the cholera outbreaks in England. He found in several cases that sewage or night soil had contaminated the drinking water in wells from which cholera cases had drawn water. No cases of cholera were found in families whose water came from uncontaminated wells. In August and September 1854, a cholera epidemic raged in London, with 500 deaths within a range of 250 yards. By careful and systematic analysis, Snow observed that the only common factor was the consumption of water from the Broad Street pump. Two pieces of evidence were telling. Firstly, a man from Brighton came to visit his brother who was ill with cholera. The brother had already died and the man stayed at his house for only 20 minutes, but while he was there he consumed a brandy and water; the next day he died of cholera. Secondly, a lady who lived in another part of London, but preferred the water from Broad Street to her local water provided additional evidence. A carrier collected water at the Broad Street pump and brought it to her house. She and her niece drank the water and died of cholera within two days. The outbreak started on 31 August, it was thought that the well was contaminated by a local cesspit that received water from a household where a baby had developed cholera on 28 August. Snow postulated that transmission of cholera was due to some "morbid material" in cholera faeces that could contaminate drinking water and reproduce in the person who drank that water (Snow, 1855).

Other studies led to the concept of the faecal indicator. In 1885, Escherich described several microorganisms in the faeces of new-born and suckling babies. This included a motile, rod-shaped microorganism that could cause milk to clot, which he named *Bacterium coli commune* (commonly referred to as *Bacterium* or *Bacillus coli*). He observed that within a few weeks after birth, this bacterium became the dominant organism in the infant colon. Other workers showed that microorganisms consistent with Escherich's description of *Bacterium coli* were invariably found in faeces. Schardinger proposed in 1892 that, since *Bacterium coli* was a characteristic component of the faecal flora, its presence in water could be taken as *"an indication of the presence of faecal pollution and therefore of the potential presence of enteric pathogens"*.

## 1.2.1 Refinement

The notion of examining microbial indicators of faecal pollution continued to be developed. Soon after the description of *Bacterium coli*, other Gram-negative, lactose-fermenting bacteria were isolated from stools and water (*Klebsiella* in 1882; *Aerobacter* [now *Enterobacter*] in 1890). Since 1901, these bacteria have been grouped under the name coliforms. The coliforms were defined as Gram-negative, non-spore-forming facultatively anaerobic bacilli that ferment lactose with production of acid and gas within 48 hours at 35°C. The definition was based on detection methods that allowed for simple isolation and enumeration of coliforms. When this methodology was applied, it soon became apparent that many genera and species that meet the coliform definition are not, or are only rarely, related to faecal contamination (Geldreich *et al.*, 1962; Mack, 1977). Under certain conditions they are also able to multiply in the aquatic environment, thus reducing their value as an indicator of faecal contamination. Already in 1904, Eijkman adopted modifications in the detection methodology that included a higher incubation temperature, which improved the specificity of the indicator. Further modifications of his method have improved the methodology for detecting these thermotolerant coliforms (also called faecal coliforms, although this is not a proper description - see Chapter 2). Although significantly more specific for faecal contamination, this parameter also had similar shortcomings. It became apparent that other bacteria (mostly *Klebsiella*), which meet the criteria for thermotolerant coliforms, originate from non-faecal environments, such as paper mill or potato-industry wastewater and other high carbohydrate wastewater (Dufour and Cabelli, 1975).

It was eventually shown that among the thermotolerant coliforms *Escherichia coli* is the preferred microbial indicator of faecal pollution (Dufour, 1977), as it is the only member of the coliform group that is invariably found in faeces of warm-blooded animals and it outnumbers the other thermotolerant coliforms in both human and animal excreta. Other microorganisms have been suggested as microbial indicators of faecal pollution (see Chapter 2), such as enterococci (previously named faecal streptococci), coliphages and suphite-reducing clostridial spores.

Although faecally derived coliforms, thermotolerant coliforms and/or *E. coli* have several drawbacks, they have historically been very useful and they are, undoubtedly, the most commonly used microbial parameters for testing drinking water quality. Their use has led to significant improvement in the safety of drinking water world-wide and they have been adopted in the World Health Organization (WHO) drinking water quality guidelines and all national drinking water quality standards. One of the main reasons for their success is the ease of the assay. In contrast with the approach to chemical contaminants of water, microbiologists soon realised the complexity that would be involved in trying to assay water for all enteric pathogens. As the common source of these pathogens was faecal pollution, microbiologists aimed for a universal microbial indicator of faecal contamination.

The ease and low cost of the assay means that it is possible to test water bodies frequently. Faecal contamination varies and it is likely that peak contamination will present the highest health risk. The importance of frequent testing has long been widely recognised:

> *"It is of the utmost importance for the control of the hygienic quality of the water supply that the bacteriological examination of both the water entering the distribution system and the water in the distribution system itself be carried out frequently and regularly" (WHO,1976);*

and

> *"It is far more important to examine a (water) supply frequently by a simple test than occasionally by a more complicated test or series of tests" (Anon, 1969).*

## 1.3    Defining the role of the indicator concept

The traditional role of indicator parameters in drinking water was as an index of faecal pollution and, therefore, likely health risk (see Box 1.2). The original microbial parameters were all bacteria that, to a greater or lesser degree, were derived from faecal contamination. Faecal-oral illness, however, is not only caused by enteric bacteria but may result from infection with pathogenic viruses or protozoa. The viruses and protozoa have different environmental behaviour and survival characteristics to bacteria, which means that faecal bacteria are not always an adequate indicator of their presence or absence. This is especially true for disinfected drinking water, as bacteria are very sensitive to disinfectants while viruses and parasites can be extremely resistant. Thus, the basic premise that the concentration of indicator organisms should be related to the extent of faecal contamination and by implication to the concentration of pathogens and the incidence of waterborne disease can not be maintained (Pipes, 1982). The roles of the indicator concept, however, are gradually expanding as is the number of possible indicator parameters. There is now a need better to define these specific roles such as in source assessment, validation of the drinking water treatment process, operational and routine monitoring as well as the traditional verification of the end product (see Chapter 1.4).

## Box 1.2. Indicator concept and criteria

Microbial indicators of pollution have been in use for decades. They were originally developed as measures of faecal pollution of source waters and subsequently the same organisms were applied to measure efficiency of treatment and post-treatment contamination and deterioration. Mossel (1978) credited Ingram with recognising the different roles to which so-called indicators were being applied and proposing that the term 'indicator' should be used for the assessment of treatment process effectiveness, while 'index' should be used for the original role of indicators, that is as a measure of faecal pollution. The search for microbial faecal indicators was based on several criteria that were well accepted by the scientific community, but were based on the assumption that the same organism would serve as both index and indicator. The criteria were:

- The indicator should be absent in unpolluted water and present when the source of pathogenic microorganisms of concern is present.
- The indicator should not multiply in the environment.
- The indicator should be present in greater numbers than the pathogenic micro-organisms.
- The indicator should respond to natural environmental conditions and water treatment processes in a manner similar to the pathogens of concern.
- The indicator should be easy to isolate, identify and enumerate.

Over time, the following criteria have been added to the original list:

- The test should be inexpensive thereby permitting numerous samples to be taken.
- The indicator should not be a pathogenic microorganism (to minimise the health risk to analysts).

The detection of pathogenic microorganisms is not normally associated with the indicator concept, as each pathogen essentially represents only itself and its absence is not an indication of the absence of other pathogens. The only current usage of a pathogen that meets the indicator concept is the detection of cryptosporidial oocysts as an indicator of treatment efficiency in the UK.

The list of microbial parameters has grown with time and these have been applied to a variety of environments, although in some instances their application strayed away from the original concept (i.e. relationship to faecal pollution), with indicators being used inappropriately.

Throughout this book, guidance is given on the best use of the various microbial and non-microbial parameters to fulfil the criteria for specific purposes. These purposes are outlined below, and in many cases may require the use of more than one microbial and/or non-microbial parameter.

- Index (or indicator) of faecal pollution in ambient waters not receiving any treatment (including water abstracted for drinking water purposes).
- Index (or indicator) of faecal pollution of groundwater.
- Indicator of treatment removal or disinfection efficiency.
- Indicator of recontamination of treated water within the distribution system.
- Models for pathogenic microorganisms.

## *1.3.1    Current practice*

The basic idea behind the use of traditional faecal indicator parameters (*i.e.* when they are absent, pathogens are absent), while not universally valid, is still applied and useful today if the parameter is chosen correctly. The most common uses are for monitoring drinking water at the tap and as it leaves the treatment works. Despite the shortcomings that have been recognised for some time, in many jurisdictions this is still done by analysing for the absence of coliforms, with or without complementary testing for *E.coli* or thermotolerant coliforms. Once the water is distributed, a positive coliform test may indicate the presence of faecal contamination but could also be derived from a non-faecal origin. Thus, the test that is used as the primary warning of faecal contamination gives very little information on the presence or absence of a health risk. Confirmation of the faecal origin is embedded in most regulations and requires testing for thermotolerant coliforms or *E. coli*. WHO (1993) indicates that *E. coli* is the parameter of choice for monitoring drinking water quality (with thermotolerant coliforms as an alternative). Enterococci and sulphite-reducing clostridia are also used as additional parameters of faecal contamination or to monitor the integrity of the distribution or storage system. Less common is their use to classify the source water, with the level of treatment to produce safe drinking water being set accordingly (more details on the use of indicator parameters for specific purposes is given in Chapter 2).

The major problem, in terms of public health protection is that (for the most part) monitoring the safety of drinking water is reactive, in the sense that any event or breakdown in the system can occur many hours and sometimes days, before it is detected by monitoring for any of the microbial parameters. This is related to both the nature of the microbial testing, which currently requires at least a day to produce a result, and also to the monitoring strategy, which has traditionally focussed on water as it leaves the treatment works and on the distribution system.

## 1.3.2    New challenges

While the use of (thermotolerant) coliforms and enterococci as indices of faecal pollution has proved successful in preventing the spread of waterborne cholera and typhoid, in the 1960s a new challenge to public health was identified. It was increasingly recognised that enteric viruses, such as hepatitis A and other enteroviruses, could also be transmitted through drinking water (Anon, 1999). Viral contamination of water also originates from pollution with human excreta, but the nature of viruses is very different from that of bacteria. They are much smaller and therefore less likely to be removed during filtration or soil passage and their resistance to disinfection is typically greater. The occurrence of outbreaks of viral illnesses associated with drinking water meeting the coliform standards indicated that coliforms were an inadequate parameter to assess the virological quality of treated drinking water (Berg and Metcalf, 1978; Petrilli *et al.*, 1974; Melnick and Gerba, 1982). Water microbiologists sought suitable alternative microbial parameters and found several groups of viruses that infect bacteria, known as bacteriophages (phages), which have a similar size and also structure characteristics to human pathogenic viruses. These were suggested as being appropriate models for the potential presence of viruses and for their survival and behaviour in the environment, as well as their removal and inactivation by water treatment and disinfection processes (Grabow *et al.*, 1984; Havelaar *et al.* 1993).

More recently, a further challenge was identified with the outbreaks of intestinal illness due to the protozoa *Giardia* sp. and *Cryptosporidium* sp. As with viruses, outbreaks have occurred without any indication, from the coliform testing, that water quality was compromised (Barrell *et al.*, 2000). It was recognised that the failure of the coliform bacteria standard was due to the more robust nature of the protozoa to disinfection, resulting in inactivation of the indicator bacteria but not the viral and protozoan pathogens. Spores of the bacterium *Clostridium perfringens* and sulphite-reducing clostridia, which are also known to be robust and resistant to disinfection have been proposed as alternative microbial parameters for such protozoa. Other indicator parameters that have been suggested to assess treatment efficiency for the removal of pathogens are aerobic spores (Chapter 2, USEPA, 2000).

As mentioned earlier, a drawback to the current use of microbial parameters, in terms of public health protection, is the reliance on end-product monitoring. End-product monitoring cannot always safeguard health but acts to verify (or not) the effectiveness of the treatment barriers. This can provide important management information (see Chapter 7) and is a useful check, which will determine any performance deficiency and also allow an assessment of any corrective procedures. Its main purpose, therefore, is to verify the efficiency of treatment and disinfection and detect post-treatment contamination.

While traditional microbial parameters have proved useful and still have an important role to play, monitoring of different aspects of the supply chain as well as possible health effects requires the use of different applications of traditional microbial parameters, different parameters and different approaches. There are two major initiatives that move to address this challenge:

- The development of water safety plans (see Box 1.3).

- The assessment of risk at all stages between catchment and consumer (Figure 1.1).

**Figure 1.1. "Catchment to consumer" approach to risk management of the safety of drinking water**

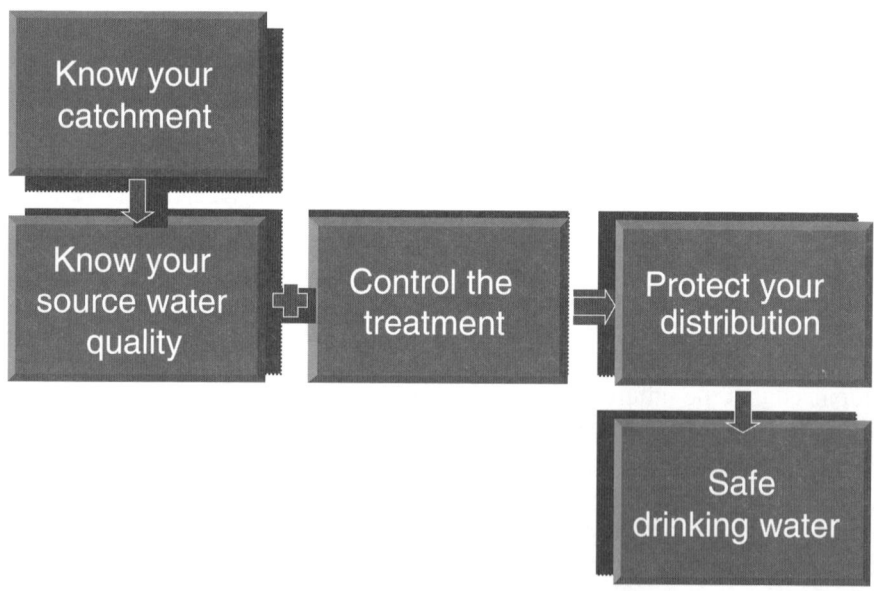

In addition, the development of rapid testing and molecular techniques for microbial parameters and pathogen detection (Chapter 1.5) may play a supporting role, particularly in risk assessment (Chapter 3) and outbreak investigation. For example, molecular techniques have, in a number of cases, allowed the identification of the source of contamination in waterborne outbreaks (Chapter 7).

## 1.4     Emergence of a new paradigm: "Due diligence"

The concept of due diligence, which means the prevention of foreseeable harm at reasonable cost, takes a significant step in changing the "reactive and sanctioning" paradigm under which suppliers (including water suppliers) operate. Demonstration of due diligence requires showing that all reasonable measures have been taken in advance to prevent the occurrence of negative health consequences. Thus, when a potentially adverse event is identified, a precautionary approach should be used. One such approach, which came out of the space programme in the 1960s, is HACCP (Hazard Analysis Critical Control Point), illustrated in Figure 1.2, which has been adapted for drinking water use and incorporated in 'Water Safety Plans' (Box 1.3 and Figure 1.3).

**Figure 1.2.  Steps in the development of a HACCP Plan**

(Adapted from Deere *et al.*, 2001)

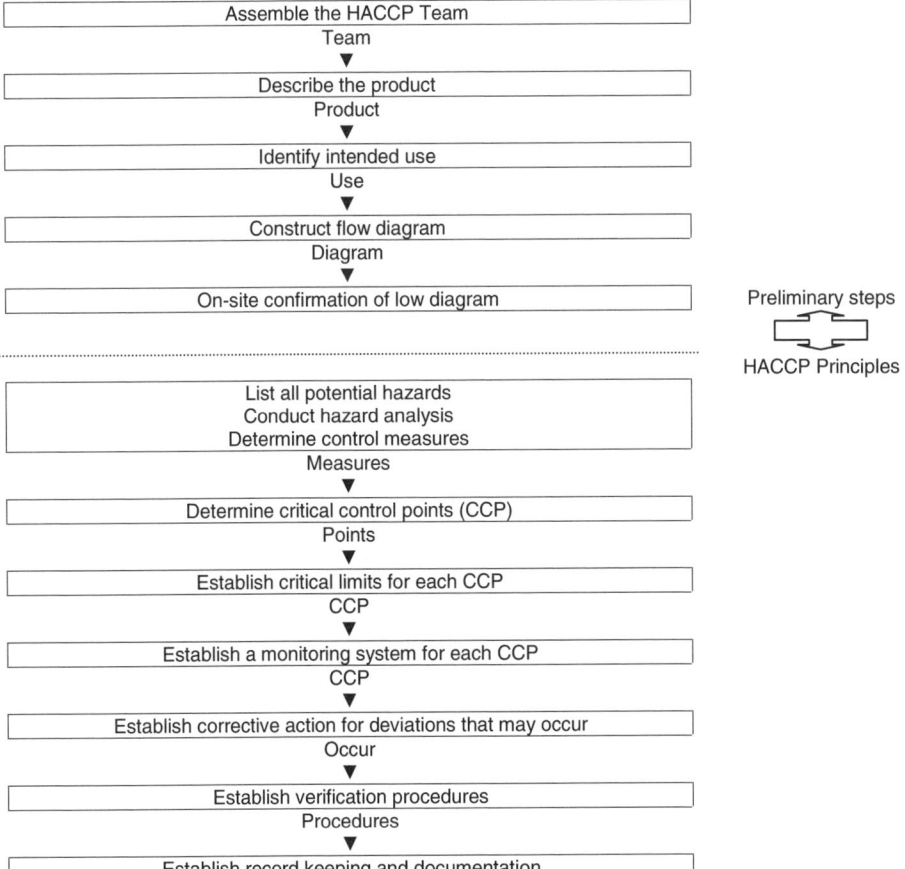

---

**Box 1.3. Water safety plans for drinking water supply**

The basis of ensuring water safety has five key components:

1.  Water quality targets based on public health protection and disease prevention.

2.  System assessment to determine whether the water supply chain (up to the point of consumption) as a whole can deliver water of a quality that meets the defined targets.

3.  Monitoring the steps in the supply chain that are of particular importance in securing safe drinking water.

4.  Management plans describing actions to be undertaken from normal conditions to extreme events.

5.  Systematic independent surveillance that verifies that the above are operating properly.

The management plans developed by water suppliers can be best termed a Water Safety Plan (WSP). The control of microbiological and chemical quality of drinking water requires the development of WSPs which, when implemented, provide the basis for process control to ensure pathogen and chemical loads are acceptable. Implicit within this process is that a tolerable disease burden has been identified at national and local levels and that water quality targets have been established for each type of technology used.

The delivery of 'safe' water, therefore, involves actions by a variety of stakeholders, as illustrated in Figure 1.3.

Water quality targets should be set to improve public health. Water suppliers have a basic responsibility to provide safe water and would be expected to undertake actions meeting points 2–4 above. The final component would normally be fulfilled by a regulatory body, which may in practice be the health, environment of local government sectors. All these processes are important in the delivery of good drinking water quality and are the subject of other texts linked to the WHO Guidelines for Drinking water Quality.

---

**Figure 1.3.  Protecting public health through ensuring drinking water quality**

(Davison *et al.*, 2002)

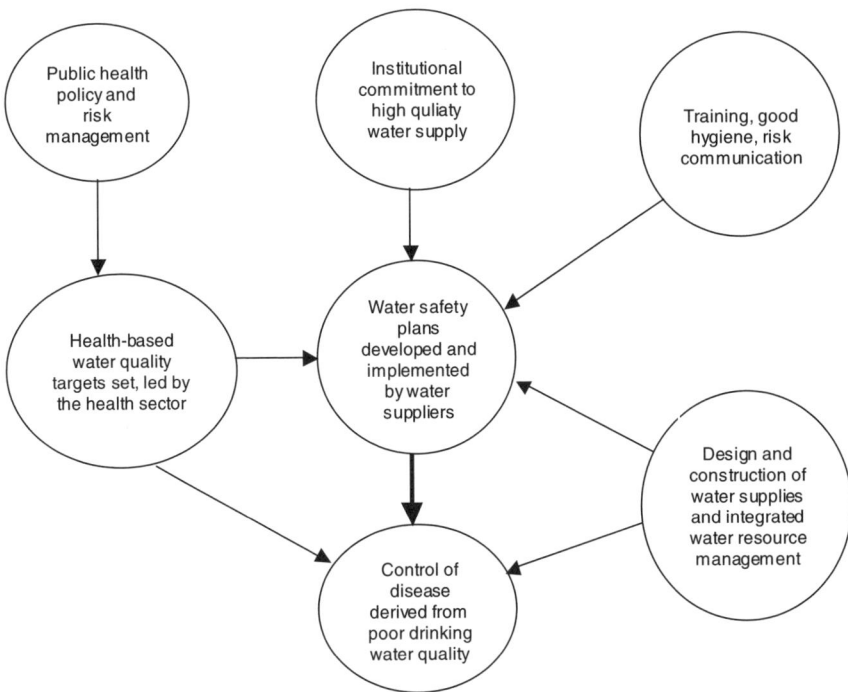

In a drinking water application, HACCP is a source-to-tap system. Microbial safety is safeguarded through knowledge of the (variations in) quality of the source water, control of the treatment process and the integrity of the distribution or storage system. No single microbial (or non-microbial) parameter is adequate to determine if all steps in this system are working properly in all circumstances. Chapters 2, 4, 5 and 6 outline which parameters are most appropriate at each stage. The resulting data can be integrated into risk assessment models (quantitative or qualitative) or be used to complement epidemiological methods, as outlined in Chapter 3.

## 1.5    Direct pathogen testing

The discovery and characterisation of a significant number of gastrointestinal pathogens led to the development of a wide array of new methods to detect and identify them (Anon, 1999; Hurst *et al.*, 2001). In addition, methods for concentrating these pathogens from natural waters were developed in the 1970s and 1980s and applied to waters potentially contaminated by faecal pollution. Results indicate that viruses (Payment and Armon, 1989; Grabow *et al.*, 2001) and parasites (LeChevallier *et al.*, 1991) may still be present in otherwise apparently safe drinking water. This, together with the recognition that illness can be associated with drinking water that meets regulatory safety standards indexed by traditional bacterial parameters (Payment *et al.*, 1991, 1997), raised questions about the safety of drinking water (LeChevallier *et al.*, 1999 a,b). The United States of America extended the principle of "pathogen-free water" to a goal of zero level of pathogens, while recognising that, in reality, this implied a judgement on the level of treatment necessary to achieve minimal or tolerable risk. To determine these treatment goals, a survey of source water was conducted using standardised methods to obtain data on pathogen occurrence. The data revealed the difficulties associated with such testing even for gathering data on occurrence (Chapter 4; Rosen and Ellis, 2000; Allen *et al.*, 2000).

In the United Kingdom, testing for *Cryptosporidium* oocysts became mandatory at some sites (established on the basis of a risk assessment). The UK approach requires continuous sampling of a proportion of treated water as it leaves the treatment works, in view of the fact that spot sampling for pathogens would be liable to miss treatment deficiencies of short duration. Data from the current monitoring programme are used to verify compliance with a treatment standard of removal of oocysts to less than one oocyst in ten litres of treated water (HMSO, 1999).

In addition to what is required by regulations, a number of water suppliers have initiated some form of direct pathogen testing (Allen *et al.*, 2000). Pathogen testing can be a useful tool for sanitary surveys of catchment areas, for setting treatment goals, for laboratory or pilot-scale demonstration of the efficacy of existing or new water treatments and for investigation of outbreaks. In general, pathogen testing helps to sharpen practices for providing safe drinking water, and detection of (viable) pathogens in drinking water is a strong trigger for remedial action. Pathogen testing can be done in addition to standard monitoring, but it is not a simple monitoring tool. The methodological constraints require that a clear idea as to what is to be achieved by the exercise should be formulated prior to undertaking pathogen testing (Havelaar, 1993) and subsequent results should be interpreted with caution. Low levels of enteric viruses and protozoan parasitic cysts have been found in drinking water in the absence of reported outbreaks (Gerba and Rose, 1990; Payment and Armon, 1989). This could be related to a lack of infection, mis-identification, acquired immunity or asymptomatic infections (Allen *et al.*, 2000; Issac-Renton *et al.*, 1994; Gerba and Rose, 1990; Payment *et al.*, 1991, 1997). Experience in Sydney (Australia) and Wyoming (USA), shed some light on the public, political, economic and legal implications of such findings even in the absence of any detectable health effects (Allen *et al.*, 2000). These cases show that unless it is known how to interpret the results and an emergency plan is in place to react to a positive result, the reaction may be inappropriate.

Methods for detecting pathogens in water are mostly still in the developmental stage (Chapter 8):

- Their sensitivity is still poor. The methods for detecting the pathogen are usually very sensitive, but because of the low level of pathogens in water, large volumes need to be analysed and as the detection methods can only effectively process small volumes a pre-concentration step must be undertaken.

- Only a few of the multitude of possible pathogens are currently detectable. Given that water may contain hundreds of different pathogens, and these may vary over time, the question of which pathogens to look for remains. Pathogen testing methods are relatively specific and will not detect all pathogens present. Molecular methods, coupled with high throughput parallel processing and bio-informatics, hold the promise of detecting a wider range of microorganisms, but are not yet practicable. One suggestion has been to search for the most significant pathogen as an indicator of treatment efficiency. The United Kingdom has adopted this approach and tests for Cryptosporidium oocysts in treated water from selected water treatment works.

- Analysis of water samples for pathogens requires a specialised laboratory, highly trained personnel and appropriate bio-safety containment. In industrialised countries few laboratories outside the clinical setting meet these requirements, and in many other countries such facilities do not exist. Pathogen testing may require growing and manipulating pathogens, thus the potential risk to analysts needs to be taken in consideration.

- Although some pathogens can be tested rapidly, most pathogen sampling and detection methods still have a time-to-(verified)-result of several days. Pathogen testing of treated water, therefore, does not escape the problems identified with end-product testing using traditional bacterial parameters, i.e. they signal that something is wrong after the problem has occurred.

These methodological limitations advocate the use of great care in the interpretation of results from pathogen testing. Any positive result may indicate water unsafe to drink and can be used to estimate the level of risk to consumers. Positive results should be used only in a well-managed, risk-based decision-making process. Negative results should always be viewed with some scepticism given the large number of possible pathogens and should not be used as an excuse for complacency.

Levels of pathogen testing may vary from simple occasional tests or planned experiments to routine monitoring of source water as well as treated water. However, if pathogen testing is included among the parameters, it is important that this is not done at the expense of essential basic monitoring. If pathogen testing is considered acceptable, where should it be done? Pathogen testing for an initial sanitary survey of source water is well accepted. After treatment at the waterworks, samples should be negative for infective pathogens, but what volume should be tested? How many samples should be taken to ensure that they are statistically representative? Familiarity with water treatment indicates that what is most to be feared is transient, short-lived failures of the system, which are difficult to detect. Given the high cost of pathogen testing – and this will not change in the near future – is the cost of routine pathogen testing justified? A good set of microbial and non-microbial parameters is probably more cost-effective.

Analysis of samples of distributed water presents a similar challenge. The objective is to detect recontamination of the water in the distribution system. How many samples should be taken, where should they be taken and for what pathogens? A good indicator of recontamination or disinfectant demand is probably more cost-effective as data can be obtained inexpensively for a large number of samples.

### 1.5.1    *Dose-response relationships for pathogens*

Determination of the effect of exposure to (different levels of) pathogenic microorganisms (*i.e.* dose-response functions) has allowed the design of a risk-based approach (Chapter 3), analogous to that taken against the risk of toxic chemicals in drinking water. Because the complete absence of pathogens in drinking water (zero risk) cannot currently be assured, this criterion has been replaced by the definition of an acceptable or tolerable risk level (Hunter and Fewtrell, 2001). Such a risk-based approach was developed by North American researchers in conjunction with the US Environmental Protection Agency (Haas, 1983; Rose and Gerba, 1991; Regli *et al.*, 1991). In this approach, a risk level of one infection per 10 000 persons per year is regarded as the acceptable maximum for pathogens in drinking water. This was based on findings that during each reported waterborne outbreak of giardiasis, at least 0.5 % of the population (*i.e.* 50 or more per 10 000 people) were infected. Because public water supplies should provide greater protection from waterborne disease, water treatment should ensure less than one case of microbial illness per year per 10 000 people as a reasonable goal (USEPA, 1989). *Giardia* was chosen as the target organism because it is more resistant to disinfection than most other pathogens (Regli *et al.*, 1993). This approach has been adopted or is being considered by a number of countries. In the Netherlands, for example, guidelines have been issued for maximum acceptable pathogen concentrations in drinking water, based on the $10^{-4}$ infection risk level.

### 1.5.2    *Molecular technologies*

Currently (for the most part) microbial parameter detection involves sampling and filtration followed by cultivation of the chosen microorganism on selective media and then colony counting or, in some cases, the demonstration of growth (*e.g.* presence-absence tests) – a process that can take 24 to 72 hours and may not pick up a number of microorganisms. The last two decades of the 20th Century, however, saw the development of molecular biology and the promise for rapid testing (less than eight hours). This resulted in techniques, such as polymerase chain reaction (PCR), for the rapid, sensitive and specific detection of index/indicator microorganisms and pathogens. In the field of health-related water microbiology, this has allowed the development of detection methods for non-culturable viruses, such as Norwalk-like viruses. While the conventional *Cryptosporidium* detection methods do not discriminate between human-pathogenic and non-pathogenic oocysts, the specificity of PCR and subsequent verification methods (hybridisation, sequencing) have allowed much more specific detection of pathogenic species or genotypes within *Cryptosporidium* (see Chapter 7). Rapidity is another characteristic of PCR and related molecular technologies (see Chapter 8). Several researchers have developed PCR techniques for the rapid detection of *E. coli* and coliforms, which make detection possible within several hours (Bej *et al.*, 1991; Fricker and Fricker, 1994).

One of the challenges for molecular methods is to assess the viability/infectivity of the detected microorganisms, as currently they detect the presence of a nucleic acid sequence, which may have originated from a dead organism or even from DNA that has not been decomposed in the aquatic environment. Culture techniques provide this information as only viable microorganisms are detected. Several techniques or combinations of techniques are now available to overcome the viability/infectivity problem. Examples are the use of inducible mRNA as target for RT-PCR (reverse transcriptase – polymerase chain reaction) or use of viability/infectivity methods to 'pre-culture' the target organism prior to PCR detection, such as the cell culture PCR method for *Cryptosporidium* and viruses (Chapter 8; Spinner and DiGiovanni, 2001), although pre-culture increases the overall assay time.

The taxonomy of microorganisms is now primarily based on genotyping, rather than on phenotypic characteristics. This genetic taxonomy allows the rapid characterisation and comparison of genotypes. This has been useful in investigations of outbreaks to determine the similarity of isolates from patients and suspected outbreak sources and in tracking the sources of contamination of a watershed or drinking water (Chapter 7; Kuhn *et al.*, 2000).

Advances in computer, chip, laser and optical technology have provided, and are providing, new opportunities for the detection and identification of microorganisms. New equipment has been used primarily in the research area, but several technologies are currently used in practice, such as flow cytometry and cell sorting, confocal laser scanning microscopy and laser scanning. Many methods are now being developed but, as yet, all are expensive as the equipment capital investment is high. One interesting development is the combination of computer chip technology and molecular microbiology, which should allow automated testing of multiple pathogens with a single DNA chip array (Chapter 8). The introduction of these technologies may allow rapid automated detection of (pathogenic or index/indicator) microorganisms in the near future.

The challenges that remain for these new methods are:

- Quantification. The quantitative aspects need to be improved as current molecular methods are, at best, only semi-quantitative.

- Infectivity. The viability and infectivity of the detected microorganisms is still uncertain.

- The "concentration issue". Detection of (especially pathogenic) micro-organisms in water requires assaying large volumes (0.1-100 litres or more), while the new technologies currently work with small volumes (0.00001-0.001 litres). This requires concentration methods that introduce recovery losses.

- The skills and novel infrastructure issue (both in personnel training and equipment). Further implementation of these technologies in practice ideally requires further simplification and also automation.

- Cost. The cost is still high and currently not amenable to frequent daily testing within the budget constraints of small water suppliers.

## 1.6 Information needs

The provision of safe drinking water is rarely the concern of a single body and in addition to the water supplier is likely to involve national, regional and/or local government, water authorities and public health authorities. In some cases, such as catchment control, there may be international involvement.

Traditionally, the main reason for monitoring water quality has been to verify whether the observed water quality is suitable for an intended use (in this case human consumption). Understanding the reasons for data collection (*i.e.* the purposes of the monitoring) helps to ensure that the data collected are appropriate for the management uses for which they are intended (Bartram and Helmer, 1996; Makela and Meybeck, 1996). The different information needs are described below.

### *1.6.1   Regulation*

Many agencies, including the WHO (WHO, 1976, 1997) identify two complementary roles in drinking water monitoring: quality control by the supplier and independent surveillance by the regulatory body. There is an evident complementarity between them, but the merging of the two is inappropriate because of the conflict of interests that would emerge. Nevertheless, in recent years experience with increased data sharing, generation of data by third parties and audit-based approaches has begun to accumulate and contributes to minimising duplication of effort. Where regulation is based upon monitoring and analysis, it does not typically mimic the requirements of the water supplier but can be proactive and can single out problems, such as the areas of distribution most likely to cause problems or known to be problematic.

The regulatory agency is ultimately responsible for determining the required level of safety of drinking water. It does so by setting water quality standards and ensuring that water suppliers meet them. While requirements vary widely from country to country, compliance is generally based on standards (such as the European Community Directive 98/83/EC; Anon, 1998; WHO, 1993) that are intended to protect public health. Standards are typically specified in terms of sampling frequency and distribution (at fixed and/or variable points) in relation to the population supplied and/or high-risk points. Parameters typically include simple physical and chemical tests (disinfectant residual, turbidity, etc.) and relatively frequent monitoring, with less frequent microbiological testing for indicators of faecal pollution (and sometimes other microbiological parameters) with specific follow-up requirements when standards are not met.

These standards clearly aim to limit the transmission of infectious disease through drinking water but they also influence the allocation of community resources to drinking water treatment. Drinking water is only one of the vectors of enteric infectious diseases. To be able to optimise the allocation of available resources to protection of drinking water, the regulator requires information on the contribution of drinking water to the overall disease burden of the population. At the next level of detail, policy makers require information about the most important threats to the safety of drinking water so that they can focus risk management options on the most relevant threats (Fewtrell and Bartram, 2001). For example, a system with a relatively high risk of post-treatment contamination should primarily focus on reducing this risk, rather than on the efficiency of source protection or on treatment.

## *1.6.2    Water supplier*

Water suppliers require information on the microbiological quality of their source water. Information on the contamination level of source water is the basis for the design of an adequate treatment system. Information on sources of pollution in the catchment area of abstraction sites gives both an indication of the level of contamination that may be expected and potential risk events (such as heavy rainfall leading to agricultural run-off). A catchment survey will also yield information on opportunities for catchment control. This may not be the domain of water suppliers, but allows them to choose between installing treatment barriers or trying to implement source protection measures. In the design phase, a catchment survey will aid in the selection of the best site for abstraction (Chapter 4).

A water supplier also needs to know the efficiency of the treatment processes in eliminating microorganisms; initially, in the design phase, to be able to design an adequate treatment system and subsequently, in the production phase, to ensure its adequate operation. In the latter phase, detailed information on the elimination of microorganisms during the different operational phases may help to optimise the efficiency of treatment processes (Chapter 5; LeChevailler and Au, 2002).

To determine if a treatment is adequate and drinking water is safe, a water supplier also needs water quality targets (Box 1.3; Fewtrell and Bartram, 2001).

In the risk-based approach, water quality targets should be derived from a tolerable risk level. Water quality targets are usually set by national authorities, which should decide on the tolerable risk level and derive the water quality targets that result in compliance with this risk level. In the risk-based approach, targets could also be a maximum concentration of a pathogen but these are generally not intended to be a measured target.

For process operation, water suppliers rely on process parameters such as coagulant dose. To ensure that treatment is eliminating microorganisms adequately every hour of the day, they need information on the relationship between the operational parameters and the elimination of microorganisms (Chapter 5; LeChevallier and Au, 2002). Finally, the company/agency that distributes the water to the consumer needs information about the water-quality changes that occur during distribution so as to be able to detect and respond to any unacceptable deterioration of water quality (see Chapter 6 and Ainsworth, 2002).

### 1.6.3    Public health agencies

In most countries, public health agencies are no longer directly responsible for the management of water supply and distribution systems. Because of this, very few public health specialists will expect to see routine water quality data on a regular basis. On the other hand, most public health surveillance will be directed at detecting cases of infection in the community. Should outbreaks be detected that implicate the water supply, review of routine water quality monitoring data will be part of the subsequent outbreak investigation (see Chapter 7). Screening of the water supply for pathogens may also be undertaken in any investigation.

The most notable pathogen-monitoring scheme for a public water system, as mentioned previously, was introduced in the UK for *Cryptosporidium*. In England and Wales, it is now a criminal offence to supply water containing $\geq 100$ *Cryptosporidium* oocysts/1 000 litres and supplies deemed to be at high risk have to be monitored continually (HMSO,1999). The standard chosen was based on operational rather than public health grounds and relating counts to public health risk has been difficult (Hunter, 2000).

With an overall responsibility towards public health, interest and indeed responsibility does not end at the quality of water leaving the water treatment works. Indeed, most public health interest may concern occurrences after the water supply enters the household and marginal or disadvantaged populations where no formal supply agency exists. Small community and particularly rural water supplies are especially problematic and are of concern in countries at all levels of socio-economic development. While factors such as parameter selection may not be very different for such areas, overall approaches to monitoring are extremely problematic and innovative approaches are required for effective action (Bartram, 1998; Chapter 6).

Investigation of events of public health concern may be triggered by disease surveillance, data from water suppliers or others undertaking monitoring or through informal means. Since the effects are delayed, such investigation presents special problems and is an area where newer analytical methods, such as those outlined in Chapter 1.5.2, may make a particular contribution. This aspect is addressed further in Chapters 7 and 8.

## 1.7 The new approach: Total System Approach to Risk Management

The combination of these developments is leading towards a risk-based approach to ensure the safety of drinking water (see Box 1.3). Traditionally, drinking water was regarded as safe when monitoring of the treated water did not show the presence of coliforms in daily samples of drinking water. In quantitative risk assessment, the safety of drinking water is demonstrated by collecting quantitative information on quality of the source water, efficiency of treatment and integrity of the distribution system. This has the benefit of providing water suppliers and relevant agencies with insight into the level of consumer protection and providing information on the strengths and weaknesses of the systems installed to protect drinking water quality. End-product monitoring remains important for long term verification of the control system.

A long-established principle in drinking water risk management is not to rely on a single barrier against pathogenic microorganisms, but to use a multiple barrier approach. This implies not only multiple barriers in water treatment, but a more encompassing approach from source to the consumer's tap (see Figure 1.1). As suggested above, in order to design an effective risk management strategy, information is required on:

- Sources of contamination in the catchment area. This would also include the relative contribution of these sources to the overall contamination level. Knowledge of the nature, location and contribution made by individual sources of contamination means that it is possible to predict peak events and to determine effective catchment control measures.

- Microbiological quality of the source water and its variation. The quality of the source water, both under average conditions and during peak events, determines the required level of treatment. Information on the level of contamination at the abstraction point can be used to design appropriate treatment systems and to design operational procedures to deal with most eventualities.

- Efficiency of the water treatment process in eliminating microorganisms and its variation. Information is required on the effectiveness of different treatment processes (as unit process and in combination with other processes) in eliminating pathogens (Haas, 1999; USEPA, 1991).

- Sources and risk of post-treatment contamination.

Such considerations are essentially the risk management components of an even larger framework or approach, where consideration is also given to tolerable risk, water quality targets and public health status, as illustrated in Figure 1.4.

**Figure 1.4. Decision-making framework**

(Adapted from Bartram *et al.*, 2001)

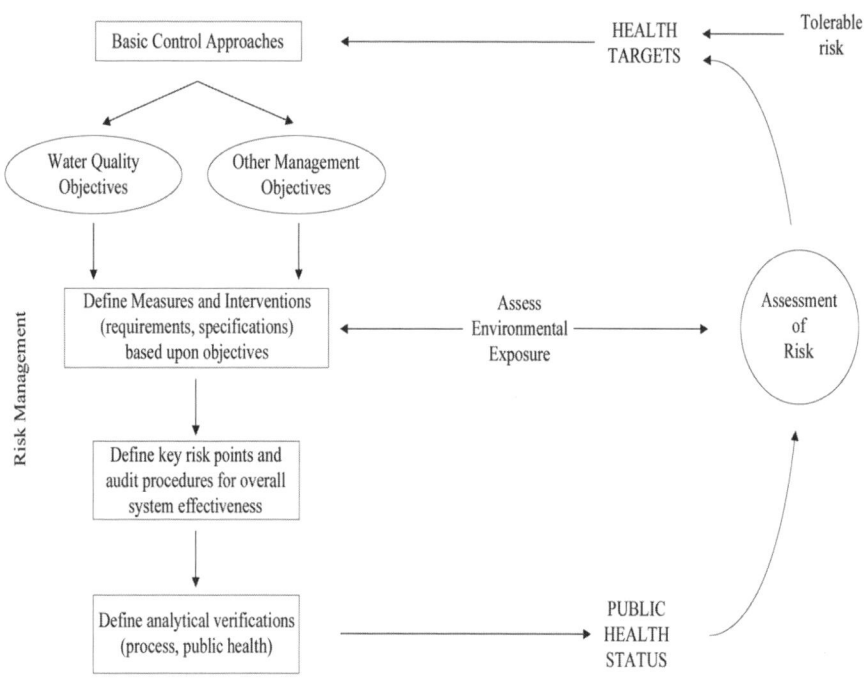

Although some index/indicator parameters can serve multiple purposes, no single parameter can fill all the information needs. Later chapters give guidance on the application of parameters for specific information needs: catchment protection, source water quality assessment, assessment of treatment efficiency, monitoring the quality of drinking water leaving the treatment facility and in the distribution system. The emphasis is on their use for demonstrating the safety of drinking water and as basis for risk management decisions.

## 1.8    Summary

Drinking water that contains pathogenic microorganisms may cause illness and, as such, it is important to have some measure (or measures) that establishes whether it is safe to drink. For the most part there are too many different pathogens to monitor and as the majority of pathogens are derived from faecal material the idea of using non pathogenic bacteria as an index of faecal pollution was developed. Initially only a few such parameters were used, but now there are more techniques and methodologies available. It is possible to monitor a wide range of index/indicator parameters (microbial and non-microbial) and also pathogens and there is a move towards using a variety of different parameters throughout the water production process and, indeed, a catchment to consumer approach to water safety plans. New methods are constantly being developed, ranging from increased pathogen detection to more real-time microbial and non-microbial parameter monitoring. The development of new and improved methodologies, along with the need for vigilance with regard to emerging hazards, results in the need for frequent re-evaluation of the best approaches and indicator parameters.

# REFERENCES

Ainsworth, R.A. (2002) Water quality changes in piped distribution systems. World Health Organization.

Allen, M.J., Clancy, J.L. and Rice, E.W. (2000) Pathogen monitoring – old baggage from the last millennium. *Journal of the American Water Works Association* **92**(9), 64-76.

Anon (1969) Reports on Public Health and Medical Subjects No. 71. Her Majesty's Stationery Office, London.

Anon (1998) European Council Directive 98/83/EC of 3 November 1998 on the quality of water intended for human consumption.

Anon (1999) *Waterborne Pathogens*. AWWA Manual of Water Practices, M48. American Water Works Association, Denver, Colorado.

Barrell, R.A.E., Hunter, P.R. and Nichols, G. (2000) Microbiological standards for water and their relationship to health risk. *Communicable Disease and Public Health* **3**(1), 8-13.

Bartram, J. and Helmer, R. (1996) Introduction. In: *Water Quality Monitoring. A practical guide to the design and implementation of freshwater quality studies and monitoring programmes.* Bartram, J. and Balance, R. (Eds.) E & FN Spon, London. pp. 1-8.

Bartram, J., Fewtrell, L. and Stenström, T-A. (2001) Harmonised assessment of risk and risk management for water-related infectious disease: an overview. In: *Water Quality: Guidelines, Standards and Health. Assessment of risk and risk management for water-related infectious disease*. Fewtrell, L. and Bartram, J. (Eds.) IWA Publishing, London. pp. 1-16

Bartram, J. (1998) Effective monitoring of small drinking water supplies. In: *Providing Safe Drinking water in Small Systems*. Cotruvo, J., Craun, G.

and Hearne, N. (Eds.) Lewis Publishers, Boca Raton, Florida. pp. 353-366.

Bej, A.K., Dicesare, J.L., Haff, L. and Atlas, R.M. (1991) Detection of *Escherichia coli* and *Shigella* spp. in water using the polymerase chain reaction and gene probes for uid. *Applied and Environmental Microbiology* **57**, 1013-1017.

Berg, G.T. and Metcalf, T. (1978) Indicators of viruses in waters. In: *Indicators of Viruses in Water and Food*. Ann Arbor Science.

Bruce-Grey-Owen Sound Health Unit (2000) The investigative report on the Walkerton outbreak of waterborne gastroenteritis. http://www.publichealthbrucegrey.on.ca/private/Report/SPReport.htm

Budd, W. (1873) *Typhoid fever: its nature, mode of spreading and prevention*. Longmans, London. 193 pp.

CDR (1998) Emerging pathogens and the drinking water supply. *CDR Weekly* **8**(33), 292.

Davison, A., Howard, G., Stevensm M., Callan, P., Kirby, R., Deere, D. and Bartram, J. (2002) *Water Safety Plans*. WHO/SHE/WSH/02/09 World Health Organization, Geneva, Switzerland.

Deere, D., Stevens, M., Davison, A., Helm, G. and Dufour, A. (2001) Management Strategies. In: *Water Qulaity: Guidelines, Standards and Health. Assessment of risk and risk management for water-related infectious disease*. Fewtrell, L. and Bartram, J. (Eds.) IWA Publishing, London. pp. 257-288.

Dufour, A.P. (1977) *Escherichia coli*: the fecal coliform. In: *Bacterial indicators/health hazards associated with water*. Hoadley, A.W. and Dutka, B.J. (Eds.) ASTM, Philadelphia. pp. 48-58.

Dufour, A.P. and Cabelli, V.J. (1975) Membrane filter procedure for enumerating the component genera of the coliform group in seawater. *Applied Microbiology* **26**, 826-833.

Ebringer, A. and Wilson, C. (2000) HLA molecules, bacteria and autoimmunity. *Journal of Medical Microbiology* **49**(4), 305-311.

Edelman, R. and Levine, M.M. (1986) Summary of an international workshop on typhoid fever. *Reviews of Infectious Disease* **8**, 329-349.

Eijkman, C. (1904) Die Gärungsprobe bei 46°C als Hilfsmittel bei der Trinkwasseruntersuchung. *Cbl. Bakteriol. Abth I. Orog.* **37**, 742-752.

Escherich, T. (1885) Die Darmbakterein des Neugeborenen und Säuglings. *Fortschritte der Medecin* **3**, 515 and 547.

Ferreira Jr., A.G., Ferreira, S.M., Gomes, M.L. and Linhares, A.C. (1995) Enteroviruses as a possible cause of myocarditis, pericarditis and dilated cardiomyopathy in Belem, Brazil. *Brazilian Journal of Medical and Biological Research* **28**(8), 869-874.

Fewtrell, L. and Bartram, J. (2001) *Water Quality: Guidelines, Standards and Health. Assessment of risk and risk management for water-related infectious disease.* IWA Publishing, London, UK.

Frankland, P. and Frankland, P. (1894) *Microorganisms in Water; Their Significance, Identification and Removal.* Longmans, Green & Co., London, UK.

Fricker, E.J. and Fricker, C.R. (1994) Application of polymerase chain reaction to the identification of *Escherichia coli* and coliforms in water. *Letters in Applied Microbiology* **19**(1), 44-46.

Geldreich, E.E., Huff, C.B., Bordner, R.H., Kabler, P.W. and Clark, H.F. (1962) The faecal coli-aerogenes flora of soils from various geographical areas. *Journal of Applied Bacteriology* **25**, 87-93.

Gerba, C.P. and Rose, J.B. (1990) Viruses in source and drinking water. In: *Drinking Water Microbiology: Progress and Recent Developments.* McFeters, G.A. (Ed.). Springer-Verlag, New York, USA.

Grabow, W.O.K., Coubrough, P., Nupen, E.M. and Bateman, B.W. (1984) Evaluation of coliphages as indicators of virological quality of sewage-polluted water. *Water SA* **10**(1), 7-14.

Grabow, W.O.K., Taylor, M.B. and de Villiers, J.C. (2001) New methods for the detection of viruses: call for review of drinking water quality standards. *Water Science and Technology* **43**(12), 1-8.

Haas, C.N. (1999) Disinfection. In: *Water Quality and Treatment: A Handbook of Community Water Supplies. Fifth Edition.* Letterman, R.D. (Ed.) McGraw-Hill, New York, USA. pp.14.1-14.6.

Haas, C.N. (1983) Estimation of risk due to low doses of microorganisms: a comparison of alternative methodologies. *American Journal of Epidemiology* **118**, 573-582.

Havelaar, A.H., van Olphen, M. and Drost, Y.C. (1993) F-specific RNA bacteriophages are adequate model organisms for enteric viruses in fresh water. *Applied and Environmental Microbiology* **59**, 2956-2962.

Havelaar, A.H. (1993) The place of microbiological monitoring in the production of safe drinking water. In: *Safety of Water Disinfection. Balancing chemical and microbial risks.* Craun G.F. (Ed.) ILSI press, Washington, DC.

HMSO (1999) The Water Supply (Water Quality) (Amendment) Regulations 199, *Statutory Instrument 1999 No. 1524*. Her Majesty's Stationery Office, London.

Horwitz, M.S., Bradley, L.M., Harbertson, J., Krahl, T., Lee, J. and Sarvetnick, N. (1998) Diabetes induced by Coxsackie virus: initiation by bystander damage and not molecular mimicry. *Nat. Med.* **4**(7), 781-785.

Hunter, P.R. and Fewtrell, L. (2001) Acceptable risk. In: *Water Quality: Guidelines, Standards and Health. Assessment of risk and risk management for water-related infectious disease.* Fewtrell, L. and Bartram, J. (Eds.) IWA Publishing, London. pp.207-227.

Hunter, P.R. and Syed, Q. (2001) Community surveys of self-reported diarrhoea can dramatically overestimate the size of outbreaks of waterborne cryptosporidiosis. *Water Science and Technology* **43**, 27-30.

Hunter, P.R. (1997) *Waterborne Disease. Epidemiology and Ecology*, John Wiley and Sons, Chichester, United Kingdom.

Hunter, P.R. (2000) Advice on the response to reports from public and environmental health to the detection of cryptosporidial oocysts in treated drinking water. *Communicable Disease and Public Health* **3**, 24-27.

Hurst, C.J., Knudsen, G.R., McInerney, M.J., Stetzenbach, L.D. and
    Walter, M.V. (2001) *Manual of Environmental Microbiology, 2nd Edition*.
    American Society for Microbiology Press, Washington, DC.

Huttly, S.R.A. (1990) The impact of inadequate sanitary conditions on health in
    developing countries. *World Health Statistics Quarterly* **43**,118-126.

Isaac-Renton, J., Moorhead, W. and Ross, A. (1996) Longitudinal studies of
    *Giardia* contamination in two adjacent community drinking water
    supplies: cyst levels, parasite viability and health impact. *Applied and
    Environmental Microbiology* **62**, 47-54.

Issac-Renton, J., Lewis, L., Ong, C. and Nulsen, M. (1994) A second
    community outbreak of waterborne giardiasis in Canada and serological
    investigation of patients. *Transactions of the Royal Society of Tropical
    Medicine and Hygiene* **88**, 395-399.

Koch, R. (1893) Ueber denaugenblicklichen stand der bakeriologischen Cholera
    diagnose. *Zeitschift für Hygiene* **XIV**, 319.

Kuhn, I., Iversen, A., Burman, L.G., Olsson-Liljequist, B., Franklin, A., Finn,
    M., Aerestrup, F., Seyfarth, A.M., Blanch, A.R., Taylor, H., Capllin, J.,
    Moreno, M.A., Dominguez, L. and Mollby, R. (2000) Epidemiology and
    ecology of enterococci, with special reference to antibiotic resistant
    strains, in animals, humans and the environment. Example of an ongoing
    project within the European research programme. *Internal Journal of
    Antimicrobial Agents* **14**(4), 337-342.

LeChevallier, M.W. and Au, K.K. (2002) Water treatment for microbial control:
    A review document. World Health Organization.

LeChevallier, M.W., Abbaszadegan, M., Camper, A.K., Hurst, C.J.,
    Izaguirre, G., Marshall, M.M., Naumovitz, D., Payment, P., Rice, E.W.,
    Rose, J., Schaub, S., Slifko, T.R., Smith, D.B., Smith, H.V.,
    Sterling, C.R. and Stewart, M. (1999a) Committee report: Emerging
    pathogens – bacteria. *Journal of the American Water Works Association*
    **91**(9),101-109.

LeChevallier, M.W., Abbaszadegan, M., Camper, A.K., Hurst, C.J.,
    Izaguirre, G., Marshall, M.M., Naumovitz, D., Payment, P., Rice, E.W.,
    Rose, J., Schaub, S., Slifko, T.R., Smith, D.B., Smith, H.V.,
    Sterling, C.R. and Stewart, M. (1999b) Committee report: Emerging

pathogens - viruses, protozoa, and algal toxins. *Journal of the American Water Works Association* **91**(9),110-121.

LeChevallier, M.W., Norton, W.D. and Lee, R.G. (1991) *Giardia* and *Cryptosporidium* spp. in filtered drinking water supplies. *Applied and Environmental Microbiology* **57**(9), 2617-2621.

Mack, W.N. (1977) Total coliform bacteria. In: *Bacterial Indicators/Health Hazards Associated with Water*. Hoadley, A.W. and Dutka, B.J. (Eds.) ASTM, Philadelphia, pp. 59-64.

MacKenzie, W.R., Hoxie, N.J., Proctor, M.E., Gradus, M.S., Blair, K.A., Peterson, D.E., Kazmierczak, J.J., Addiss, D.G., Fox, K.R., Rose, J.B. and Davis, J.P. (1994) A massive outbreak in Milwaukee of *Cryptosporidium* infection transmitted through the public water supply. *New England Journal of Medicine* **331**(3), 161-167.

Makela, A. and Maybeck, M. (1996) Designing a monitoring programme. In: *Water Quality Monitoring*. Bartram, J. and Balance, R. (Eds.) E&FN Spon, London, pp. 35-59.

McFeters, G.A. (1990) *Drinking Water Microbiology*. Springer-Verlag, New York.

Mead, P.S., Slutsker, L., Dietz, V., McCraig, L.F., Bresee, J.S., Shapiro, C., Griffin, P.M. and Tauxe, R.V. (1999) Food-related illness and death in the United States. *Emerging Infectious Diseases* **5**(5), 607-625.

Melnick, J.L. and Gerba, C.P. (1982) Viruses in surface and drinking waters. *Environmental International* **7**, 3-7.

Morris, R.D. and Levine, R. (1995) Estimating the incidence of waterborne infectious disease related to drinking water in the United States. In: *Assessing and Managing Health Risks from Drinking Water Contamination: Approaches and Applications*. Reichard, E.G., Zapponie, G.A. (Eds.) IAHS Press, Wallingford, Oxfordshire, United Kingdom, pp. 75-88.

Mossel, D.A.A. (1978) Index and indicator organisms: a current assessment of their usefulness and significance. *Food Technology, Australia* **30**, 212-219.

Payment, P. (1997) Epidemiology of endemic gastrointestinal and respiratory diseases – incidence, fraction attributable to tap water and costs to society. *Water Science and Technology* **35**, 7-10.

Payment, P. and Armon, R. (1989) Virus removal by drinking water treatment processes. *CRC Critical Reviews in Environmental Control* **19**, 15-31.

Payment, P., Richardson, L., Siemiatycki, J., Dewar, R., Edwards, M. and Franco, E. (1991) A randomized trial to evaluate the risk of gastrointestinal disease due to consumption of drinking water meeting currently accepted microbiological standards. *American Journal of Public Health* **81**, 703-708.

Payment, P., Siemiatycki, J., Richardson, L., Renaud, G., Franco, E. and Prévost, M. (1997) A prospective epidemiological study of gastrointestinal health effects due to the consumption of drinking water. *International Journal of Environmental Health Research* **7**, 5-31.

Petrilli, F.L., Crovari, P., DeFlora, S. and Vannucci, A. (1974) The virological monitoring of water. I. Drinking water. *Boll. Ist. Seiroter, Milan* **53**, 434-442.

Pipes, W.O. (1982) Indicators and water quality. In: *Bacterial Indicators of Pollution*. Pipes W.O. (Ed.). CRC Press, Boca Raton. pp. 83-96.

Prendergast, M.M. and Moran, A.P. (2000) Lipopolysaccharides in the development of the Guillain-Barré syndrome and Miller Fisher syndrome forms of acute inflammatory peripheral neuropathies. *Journal of Endotoxin Research* **6**(5), 341-359.

Prüss, A., Kay, D., Fewtrell, L. and Bartram, J. (2002) Estimating the burden of disease due to water, sanitation and hygiene at global level. *Environmental Health Perspectives* IN PRESS.

Regli, S., Berger, P. and Macler, B. (1993) Proposed decision tree for management of risks in drinking water: consideration for health and socioeconomic factors. In: *Safety of Water Disinfection: Balancing Chemical and Microbial Risks*. Craun G.F. (Ed.) ILSI Press, Washington, D.C. pp. 39-80.

Regli, S., Rose, J.B., Haas, C.N. and Gerba, C.P. (1991) Modelling the risk from *Giardia* and viruses in drinking water. *Journal of the American Water Works Association* **83**(11), 76-84.

Roivainen, M., Rasilainen, S., Ylipaasto, P., Nissinen, R., Ustinov, J., Bouwens, L., Eizirik, D.L., Hovi, T. and Otonkoski, T. (2000) Mechanisms of coxsackievirus-induced damage to human pancreatic beta-cells. *Journal of Clinical Endocrinology and Metabolism* **85**(1), 432-440.

Rose, J.B. and Gerba, C. (1991) Use of risk assessment for development of microbial standards. *Water Science and Technology* **24**(2), 29-34.

Rosen, J. and Ellis, B. (2000) The bottom line on the ICR Microbial data. Paper ST6-3 In: *Proceedings of AWWA Water Quality Technology Conference 2000*. Salt Lake City, Utah.

Schardinger, F. (1892) Ueber das Vorkommen Gahrung Errengeneder Spaltpilze im drinkwasser und ihre Bedeutung for die Hygienishe Beurthelung Desselben. *Wien. Klin. Wochschr.* **5**, 403-405.

Shanmugam, J., Raveendranath, M. and Balakrishnan, K.G. (1986) Isolation of ECHO virus type-22 from a child with acute myopericarditis – a case report. *Indian Heart Journal* **38**(1), 79-80.

Snow, J. (1855) *On the Mode of Communication of Cholera*. John Churchill, London.

Spinner, M.L. and DiGiovanni, G.D. (2001) Detection and identification of mammalian reoviruses in surface water by combined cell culture and reverse transcription – PCR. *Applied and Environmental Microbiology* **67**(7), 3016-3020.

Stenström, T.A. (1994) A review of waterborne outbreaks of gastroenteritis in Scandinavia. In: *Water and Public Health*. Golding, A.M.B., Noah, N. and Stanwell-Smith, R. (Eds.) Smith-Gordon & Co., London. pp. 137-143.

Uemura, N., Okamoto, S., Yamamoto, S., Matsumura, N., Yamaguchi, S., Yamakido, M., Taniyama, K., Sasaki, N. and Schlemper, R.J. (2001) *Helicobacter pylori* infection and the development of gastric cancer. *New England Journal of Medicine* **345**(11), 784-789.

US Department of Health and Human Services (1998) *Preventing emerging infectious diseases: A strategy for the 21$^{st}$ century*. US Department of Health and Human Services, Atlanta, Georgia.

USEPA (1991) *Guidance Manual for Compliance with Filtration and Disinfection Requirements for Public Water Systems using Surface Water Sources.* US Environmental Protection Agency, Washington, DC.

USEPA (1989) National Primary Drinking Water Regulations: filtration, disinfection, turbidity, Giardia lamblia, viruses, Legionella, and heterotrophic bacteria; Final Rule (40 CFR Parts 141 and 142). *Federal Register* **54**(124).

USEPA (2000) National Primary Drinking Water Regulations: Long term I enhanced surface water treatment and filter backwash rule (40 CFR Parts 141 and 142). *Federal Register Proposed Rule* **65**(69).

van der Kooij (1993) Importance and assessment of the biological stability of drinking water in the Netherlands. In: *Safety of Water Disinfection: Balancing Chemical and Microbial Risks.* Craun, G.F. (Ed.) ILSI Press, Washington, DC. pp. 165-179.

Waite, W.M. (1991) Drinking water standards – a personal perspective. In: *Proceedings of the UK Symposium on Health-related Water Microbiology.* Morris R. et al. (Eds.). International Association for Water Pollution Research and Control, London. pp 52-65.

White, G.C. (1999) *Handbook of Chlorination.* John Wiley & Sons Inc. New York.

Whitlock, E.A. (1954) The waterborne diseases of microbiological origin. Paper presented at the Annual Conf. of the Nat. Assoc. of Bath Superintendents 1954, Anderson Ltd, Stepney Green, United Kingdom.

WHO (1976) *Surveillance of Drinking water Quality*, World Health Organization, Geneva.

WHO (1993) *Guidelines for Drinking water Quality. Volume :, Recommendations. Second Edition.* World Health Organization, Geneva. Second Edition.

WHO (1997) *Guidelines for Drinking water Quality, Volume 3: Surveillance and Control of Community Supplies. Second Edition.* World Health Organization, Geneva. Second Edition

WHO (1998) Emerging and re-emerging infectious diseases. Fact sheet no. 97. (http://www.who.int/inf-fs/en/fact097.html).

*Chapter 2*

# INTRODUCING PARAMETERS FOR THE ASSESSMENT OF DRINKING WATER QUALITY

*P. Payment, M. Waite and A. Dufour*

Note: Inspiration and some text from World Health Organization (WHO) Guidelines (WHO, 1996; 1997) have been used for the preparation of this chapter.

## 2.1    Introduction

Chapter 1 introduced the index and indicator concept and outlined a number of microorganisms (and groups of microorganisms) that have been adopted in the quest to determine whether or not drinking water is microbiologically safe to drink. This chapter examines the range of both microbial and non-microbial parameters and briefly outlines their uses and applications. It is intended to act as an introduction to the parameters that can be used to assess drinking water quality and further details on their use in specific areas can be found in subsequent chapters.

The early impetus behind the bacteriological examination of drinking water was to determine whether water as consumed was contaminated. Much water consumed received no treatment and such treatment as was applied was mainly intended to improve aesthetic quality. At that time what was required was what is now referred to as an Index organism, although the term Indicator was generally applied. It has since been recognised that microbial parameters can provide useful information throughout the drinking water production process, including catchment survey, source water characterisation, treatment efficiency and examination of the distribution system. Adopting the index and indicator terminology as advocated by Waite (1991) and briefly outlined in Chapter 1, index organisms can give a measure of the amount of faecal pollution in a water source, whereas indicator parameters may be used to give information on the effectiveness with which specific groups of microorganisms have been removed or inactivated by treatment processes, with their presence after treatment indicating that pathogens may still be present. For example, the presence of spores of sulphite-reducing clostridia or bacteriophages in treated drinking water suggests that highly persistent microorganisms may have survived, while colony counts of aerobic heterotrophic bacteria or direct microscopic counts can provide information on the availability of nutrients in the water, which may result in aesthetic problems or in the presence of opportunistic pathogens.

Although many waterborne pathogens can now be detected (and, indeed, a number are outlined in this chapter) the methods for their detection are often difficult to implement, relatively expensive, and time-consuming. Furthermore, the original logic behind the indicator (now index) concept still holds true, in that a range of pathogens may be shed into water from the faecal matter of infected people and animals, and there are enteric pathogens as yet unrecognised. As such, it is neither practicable nor recommended to examine water for every known pathogen that might be present. Examination of finished waters for pathogens will only permit confirmation that consumers have been exposed to the pathogens whereas examination for non-pathogenic organisms as an index of faecal pollution or an indicator of adequacy of treatment permits recognition of the potential for pathogens to be present without the need for their actual presence. This chapter describes the index/indicator parameters and highlights those that are best suited to a range of purposes (which are explored further in subsequent chapters), with the main thrust being towards minimising faecal-oral disease transmission.

## 2.2     Microbial parameters

This section outlines microbial parameters used to assess drinking water quality, examining the most appropriate uses, ease of analysis and some of the implications and responses relating to finding a positive sample. Characteristics such as speed of measurement, technical difficulty of the assay, microbial environmental survival and resistance to treatment are summarised at the end of the Section in Table 2.1, while Table 2.2 summarises the applicability and suitability of each parameter for assessing source water, treatment efficiency and so on.

A number of documents contain detailed information on taking samples for analysis of microbial parameters and their storage and transportation (WHO, 1997; Anon, 1994; APHA, AWWA, WEF, 1998), however, there are several key points that are summarised below:

48

- Care should be taken that the samples are representative of the water examined. This has implications in terms of the location and construction of the sampling points, the frequency of sampling and also the aseptic technique employed by the sampler.

- If the sample contains disinfectant (such as chlorine, chloramine, chlorine dioxide or ozone) sterile sodium thiosulphate should be included in the sample container in order to neutralise any residual. The concentration of the residual disinfectant and the pH at the sampling point should be determined at the time of collection.

- In order to minimise changes in the microbial content, samples should not be exposed to light and should be rapidly cooled to between 4-10°C. WHO and UNEP recommend that if samples can not be cooled they should be examined within two hours of sampling (Bartram and Ballance, 1996). Examination of cooled samples should begin as soon as possible after collection, ideally within six hours, with 24 hours being considered the absolute maximum (WHO, 1997; Bartram and Ballance, 1996).

Further details on sampling can be found in Chapter 6.

Internationally accepted methods of analysis for the microbial parameters discussed in this chapter can be found in a number of sources, including Anon (1994) and APHA, AWWA, WEF (1998). The International Organization for Standardization (ISO) also prepares and publishes methods (see Chapter 8).

Most of the microbial parameters discussed below are common in the environment and can easily be introduced in the course of sampling or analysis. It is therefore advisable to be cautious in the response to their detection in a single sample of treated water in the absence of supporting factors such as treatment problems, risks pointing to recontamination in distribution or lack of residual chlorine. Their detection in the presence of supporting factors, in associated samples, or on re-sampling, however should be taken as strong evidence that the quality of the water in supply has been compromised.

### 2.2.1 *The coliform group*

The coliform group is made up of bacteria with defined biochemical and growth characteristics that are used to identify bacteria that are more or less related to faecal contaminants. The total coliforms represent the whole group, and are bacteria that multiply at 37°C. The thermotolerant coliforms are bacteria that can grow at a higher temperature (44.2°C) and *Escherichia coli* is a thermotolerant species that is specifically of faecal origin.

A finding of any coliform bacteria, whether thermotolerant or not, in water leaving the treatment works requires immediate investigation and corrective action. There is no difference in the significance of total coliforms, thermotolerant coliforms and *E. coli* in water leaving a treatment works, as they all indicate inadequate treatment, and action should not be delayed pending the determination of which type of coliform has been detected. Upon detection in a distribution system, investigations must be initiated immediately to discover the source of the contamination.

*Total coliforms:* Coliform organisms, better referred to as total coliforms to avoid confusion with others in the group, are not an index of faecal pollution or of health risk, but can provide basic information on source water quality. Total coliforms have long been utilised as a microbial measure of drinking water quality, largely because they are easy to detect and enumerate in water.

They have traditionally been defined by reference to the method used for the group's enumeration and hence there have been many variations dependent on the method of culture. In general, definitions have been based around the following characteristics: Gram-negative, non-spore-forming rod-shaped bacteria capable of growth in the presence of bile salts or other surface-active agents with similar growth-inhibiting properties, oxidase-negative, fermenting lactose at 35-37°C with the production of acid, gas, and aldehyde within 24-48 hours. These definitions presume the use of cultural methods for identification and enumeration. There has recently been a move towards a genotypic definition based on the recognition that in order to ferment lactose, organisms must possess β-galactosidase activity. Using this approach total coliforms are defined as members of a genus or species within the family *Enterobacteriaceae* capable of growth at 37°C and possessing β-galactosidase.

Traditionally, total coliforms were regarded as belonging to the genera *Escherichia, Citrobacter, Enterobacter,* and *Klebsiella.* However, regardless of the definition adopted, the group is heterogeneous. It includes many lactose-fermenting bacteria, such as *Enterobacter cloacae* and *Citrobacter freundii,* which can be found in both faeces and the environment (nutrient-rich waters, soil, decaying plant material) as well as in drinking water containing relatively high concentrations of nutrients. It also includes members of genera such as *Budvicia* and *Rahnella*, which are never found in mammalian faeces.

Because total coliforms of non-faecal origin can exist in natural waters, their presence can occasionally be tolerated in unpiped or untreated water, in the absence of more specific index parameters. Where it can be demonstrated that coliforms in water are not faecally derived and are, thus, of no sanitary significance, expenditure to achieve their eradication may be considered unnecessary and many standards require only absence of total coliforms from 95% of samples from within distribution systems. However, if used as an indicator of treatment efficiency, total coliform bacteria should not be detectable in water leaving a treatment works and in such cases their detection should provoke immediate investigation and corrective action.

They are detectable by simple, inexpensive cultural methods that require basic routine bacteriology laboratory facilities, but well-trained and competent laboratory workers. They pose very little risk to the health of laboratory workers given good standards of laboratory hygiene.

***Thermotolerant ('faecal') coliforms:*** The term 'faecal coliforms', although frequently employed, is not correct: the correct terminology for these organisms is 'thermotolerant coliforms'. Thermotolerant coliforms are defined as the group of total coliforms that are able to ferment lactose at 44-45°C. They comprise the genus *Escherichia* and, to a lesser extent, species of *Klebsiella, Enterobacter,* and *Citrobacter.* Of these organisms, only *E. coli* (covered in the next section) is considered to be specifically of faecal origin, being always present in the faeces of humans, other mammals, and birds in large numbers and rarely, if ever, found in water or soil in temperate climates that has not been subject to faecal pollution (although there is the possibility of regrowth in hot environments, Fujioka *et al.*, 1999).

Thermotolerant coliforms other than *E. coli* may originate from organically enriched water such as industrial effluents or from decaying plant materials and soils. In tropical and subtropical waters, thermotolerant coliform bacteria may occur without any obvious relation to human pollution and have been found on vegetation in the tropical rainforest. This means that the occurrence of the thermotolerant coliform group in subtropical or tropical waters or those enriched with organic wastes does not necessarily suggest faecal contamination by humans. However, their presence in treated waters should not be ignored, as the basic assumptions that pathogens may be present and that treatment has been inadequate still hold good.

Thermotolerant coliforms are a less reliable index of faecal contamination than *E. coli* although, under most circumstances and especially in temperate areas, in surface water their concentrations are directly related to *E. coli* concentrations. Their use for water-quality examination is therefore considered acceptable when no other method is available. However, as methods for the simultaneous detection of thermotolerant coliforms and of *E. coli* are available, these methods should be preferred.

51

Thermotolerant coliforms are easily detectable and a variety of internationally standardised methods and media for their detection are available (ISO 9308-1: ISO 9308-2). These methods require basic routine bacteriology laboratory facilities and well-trained and competent laboratory workers. They should pose very little risk to the health of laboratory workers given good standards of laboratory hygiene.

***Escherichia coli:*** *Escherichia coli* is a taxonomically well defined member of the family *Enterobacteriaceae*, and is characterised by possession of the enzymes β-galactosidase and β-glucuronidase. It grows at 44-45°C on complex media, ferments lactose and mannitol with the production of acid and gas, and produces indole from tryptophan. However, some strains can grow at 37°C but not at 44-45°C, and some do not produce gas. *E. coli* does not produce oxidase or hydrolyse urea. Complete identification of the organism is too complicated for routine use, but a number of tests have been developed for rapid and reliable identification with an acceptable degree of accuracy. Some of these methods have been standardised at international and national levels (*e.g.* ISO 9308-1; ISO 9308-2) and accepted for routine use, others are still being developed or evaluated.

*E. coli is* abundant in human and animal faeces, and in fresh faeces it may attain concentrations of $10^9$ per gram. It is found in sewage, treated effluents, and all natural waters and soils subject to recent faecal contamination, whether from humans, wild animals, or agricultural activity. It has been suggested that *E. coli* may be present or even multiply in tropical waters not subject to human faecal pollution (Fujioka *et al.*, 1999). However, even in the remotest regions, faecal contamination by wild animals, including birds, can never be excluded and this suggestion requires further investigation. Because animals can transmit pathogens that are infective in humans, the presence of *E. coli* must not be ignored, because, as with the presence of thermotolerant coliforms, the presumption remains that the water has been faecally contaminated and that treatment has been ineffective.

*E. coli* is widely preferred as an index of faecal contamination. It is also widely used as an indicator of treatment effectiveness although, as with the other coliform indicators, it is more sensitive to disinfection than many pathogens (in particular viruses and protozoa). The detection of *E. coli* in water leaving a treatment works is of the same significance as any other coliform organism, but its absence does not necessarily indicate that pathogens have been eliminated.

Because *E. coli* is indicative of recent faecal contamination, with any positive finding consideration should be given to whether steps need to be taken to protect consumers. In the event of more than one related sample containing *E. coli*, or the recognition of other significant features such as treatment aberrations, the issue of advice to boil water intended for drinking may be considered appropriate (see Chapter 7). However, in many instances it may be acceptable to restrict the response to the taking of additional samples and sanitary inspection in order to assist interpretation of the initial results. If the water is a treated piped supply, a positive sample suggests that a failure or ingress has occurred, such as a breakdown in disinfection, treatment before disinfection has failed, or contaminated water has entered the system. Immediate action must, therefore, be taken to discover the source of contamination and to take appropriate steps (which will depend on the level of contamination) to protect consumers until the problem is resolved.

*E. coli* is detectable by simple, inexpensive cultural methods that require basic routine bacteriology laboratory facilities, but require well-trained and competent laboratory workers. It can pose a health risk for laboratory workers as some strains of this organism are pathogenic.

## 2.2.2    *Enterococci and faecal streptococci*

Chain forming gram-positive cocci used to be placed in the genus *Streptococcus* and faecal streptococci were those streptococci generally present in the faeces of humans and animals. All possess the Lancefield group D antigen. A sub-group of the faecal streptococci, which is relatively tolerant of sodium chloride and alkaline pH, have been grouped under the genus *Enterococcus*. Most of the *Enterococcus* species are of faecal origin and can generally be regarded as specific indices of human faecal pollution for most practical purposes.

Faecal streptococci are more resistant to stress and chlorination than *E. coli* and the other coliform bacteria. Although both faecal streptococci and enterococci remain in use as monitoring parameters in drinking water, enterococci appear likely to supplant faecal streptococci as the parameter of choice as they are clearly of faecal origin from warm blooded animals. Enterococci, as an index of faecal pollution, can also be used to complement *E. coli* in catchment assessment, in tropical climates (where *E. coli* is less appropriate because of the suspicion of multiplication) and in ground water source evaluation. Enterococci can also serve as an additional indicator of treatment efficiency. They are highly resistant to drying and thus may be valuable for routine control after new mains are laid or distribution systems are repaired, or for detecting pollution of groundwater or surface waters by surface run-off. In the UK they have been used to assess the significance of doubtful results from other organisms (Gleeson and Gray, 1997).

Enterococci are detectable by simple, inexpensive cultural methods that require basic routine bacteriology laboratory facilities, but require well-trained and competent laboratory workers. They could pose a health risk for laboratory workers as some strains of these bacteria are pathogenic.

### 2.2.3    Ratios of counts

The ratio of counts of thermotolerant coliforms and faecal streptococci has been proposed as a means of differentiating between contamination from human and animal sources. Ratios of thermotolerant coliforms to faecal streptococci greater than four have been suggested to indicate a human source whereas ratios less than 0.7 indicate an animal source. These ratios are highly variable. They may vary according to the number of sources, are often site specific, differ with the effects of wastewater disinfection and the age of the contamination (due to the different survival rates of different *Enterococcus* species). All of these factors have a marked effect on the ratios. This ratio is therefore no longer recommended as a means of differentiating sources of pollution. The same applies to most ratios obtained for index, indicator and pathogenic microorganisms.

### 2.2.4    Direct total counts and activity tests (total and viable bacteria)

Quantification of the total numbers, viability or activity of microorganisms can be useful in assessing the general microbial content of water, its general cleanliness, the integrity of distribution systems and so on. However, these methods generally have little direct sanitary significance. Most direct tests are targeted to general microbial populations rather than faecal microorganisms.

Direct counts of bacteria can provide basic information on the numbers of bacteria in water during abstraction and treatment. Using vital stains, the viability of individual organisms can be assessed. More complex techniques can be used to provide information on serotype and genetic content. Very large numbers of aerobic and anaerobic bacteria are present in water and only a very small proportion can be grown on artificial media such that direct assays can be considered more representative.

Microscopic tests are performed by filtration on membrane filters and bacteria are stained with vital or non-vital stains. The tests require a very good microscope, but are not difficult to perform and can be done at relatively low cost. However, the limited sanitary significance of the results mean that these tests are generally only employed as part of research studies. Automated scanning instruments and flow cytometers can be used to determine total and viable counts more rapidly than by manual microscopy (see Chapter 8). These methods, however, are more expensive and complex than simple microscopic methods.

Assays for microbial metabolism can also be employed to assess general microbial levels. These can use sensitive chemical measures such as the determination of adenosine triphosphate (ATP – a high-energy phosphate carrier found in all living organisms) and are used for assessing microbial levels in food and pharmaceuticals. Being simple and rapid they could potentially be used in the testing of water. However, as they assess general microbial level and not faecal contamination they are of limited sanitary significance and, as such, are not used in routine monitoring.

### 2.2.5    *Heterotrophic aerobic and aerobic spore-former bacterial counts*

Colony counts of heterotrophic aerobic bacteria (often referred to as heterotrophic plate counts – HPC) and aerobic spore-former (mainly *Bacillus spp.*) bacteria can be used to assess the general bacterial content of water. They do not represent all the bacteria present in the water, only those able to grow and produce visible colonies on the media used and under the prescribed conditions of temperature and time of incubation. Colony counts are generally determined following incubation at 22°C and 37°C to assess bacteria that may be unrelated to faecal pollution. They are of little sanitary significance, but may be useful in the long-term assessment of the efficiency of water treatment, specifically the processes of coagulation, filtration, and disinfection, where the objective is to keep counts as low as possible. While actual counts are of limited value, changes from counts normally found at particular locations may warn of significant developments. They may also be used to assess the cleanliness and integrity of the distribution system and the suitability of the water for use in the manufacture of food and drink products, where high counts may lead to spoilage.

Cultural methods used for counting heterotrophic aerobic bacteria can be adapted to count only spores by exposing samples to temperatures of 70–80 °C for ten minutes before culturing. Counts of aerobic spore-forming bacteria before and after a treatment are useful in evaluating treatment effectiveness, whether removal or disinfection. They have been proposed as surrogates for the removal of cysts of parasitic protozoa but their value in this respect is as yet unproven.

Heterotrophic bacterial counts are provided by simple, inexpensive cultural methods that require basic routine bacteriology laboratory facilities and can be performed by relatively unskilled persons. They are not an index of faecal pollution but provide basic information on culturable bacteria and their viability. They are not generally considered to be a health risk for laboratory workers, although certain HPC organisms are thought to be opportunistic pathogens.

## 2.2.6    *Bacteriophages*

Bacteriophages (also known simply as phages) are viruses that only infect bacteria. Some bacteriophages are comparable in size and behaviour to human enteric viruses and they are relatively easy to detect and enumerate (see Chapter 8). Various groups and types of bacteriophage, particularly those of coliform bacteria (coliphages) and those of *Bacteroides* spp., have been proposed as indices of faecal pollution (and possible enteric virus presence) and as indicators of treatment efficiency for both water and wastewater-treatment processes. Leclerc (2000) has reviewed the literature on the use of bacteriophages and concludes that they have significant limitations as indices of faecal pollution and enteric viruses. However, other published evidence indicates that bacteriophages have potential value as indices of faecal contamination and indicators of treatment effectiveness (Sobsey *et al.*, 1995; Grabow, 2001).

*Coliphages*: Coliphages are divided here into two groups, both of which occur in sewage and faecally polluted water, where they generally outnumber human viruses. However, the frequency of occurrence of coliphages in human and animal faeces varies, and sometimes they are detected in faeces at only low frequencies. In this respect, coliphages differ from bacterial indices of faecal contamination.

- Somatic coliphages. These infect host strains via cell wall (somatic) receptors and are frequently detected in human and animal faeces. The host normally used is E.coli. The bacteriophages (coliphages) detected by currently used E. coli hosts are relatively host-specific and most coliphage isolates do not infect other bacterial species, including species that may occur naturally in the aqueous environment. It is possible, but unlikely, that somatic coliphages occur unrelated to faecal pollution. However, their

56

usefulness as an index of faecal pollution and enteric viruses is limited by inadequate knowledge of their natural history. They may, when present in raw waters, be a suitable index of faecal contamination and an indicator of virus inactivation and removal during treatment.

- F-specific RNA bacteriophages (male-specific coliphages). These infect bacteria through the F- or sex-pili. Although they are only present in the faeces of a small proportion of people, they are commonly found in high numbers in sewage. They have been used primarily as an index of sewage contamination and, because of their relatively high persistence and similarity to viruses, as an additional indicator of treatment efficiency or for groundwater protection. There are two groups of F-specific coliphages, those containing RNA and those containing DNA and both groups are found in human and animal faecal wastes. The F-specific RNA coliphages are similar in size, shape and basic composition to many human enteric viruses (single-stranded RNA surrounded by a protein coat) such as astroviruses, caliciviruses and hepatitis A and E viruses. There are four major subgroups of F-specific RNA coliphages. Because there is some evidence that the occurrence of these groups differs between humans and other animals, it may be possible to distinguish human from animal contamination by grouping the F-specific RNA coliphages isolated from faecally contaminated waters (Hsu et al., 1995).

*Bacteroides phages*: *Bacteroides* spp. outnumber the coliform group in human faeces (Gleeson and Gray, 1997), with *Bacteroides fragilis* being the most commonly found species. They are strict anaerobes and they have not been shown to multiply in the environment. Bacteriophages of *Bacteroides* have been proposed as an index of faecal pollution as they are considered to be more resistant to natural inactivation and water treatment processes than bacterial indicators and have a decay rate similar to that of human enteric viruses. The draw-backs, however, are that their densities in raw waters may be low (requiring concentration from large volumes) and the methods of detecting them in water are currently not very reliable.

Coliphages are detectable by simple, inexpensive and rapid methods that can be applied in a basic routine bacteriology laboratory. *Bacteroides* bacteriophages, however, require facilities for anaerobic culture and require a greater degree of expertise and laboratory resources. Some internationally standardised methods exist (*e.g.* ISO 10705-1; 10705-2; 10705-4). They are generally not considered to be a health risk for laboratory workers, although some of the host bacterial strains may be opportunistic pathogens.

### 2.2.7    *Sulphite-reducing clostridia and* Clostridium perfringens

Sulphite-reducing clostridia are obligately anaerobic, spore-forming organisms, of which the most characteristic, *Clostridium perfringens*, is normally present in faeces (although in much smaller numbers than *E. coli*). Except for *Clostridium perfringens* they are not exclusively of faecal origin and can be derived from other environmental sources. The spores can survive in water for very long periods and are quite resistant to disinfection. As *C. perfringens* is faecally specific, unlike the other sulphite-reducing clostridia, it is the preferred parameter. Clostridia are not, however, recommended for the routine monitoring of distribution systems because of their length of survival they may be detected long after (and far from) the pollution event, leading to possible false alarms.

The presence of *C. perfringens* in groundwaters in the absence of *E.coli* and enterococci points to pollution at some time in the past and suggests the source may be liable to intermittent contamination. Being relatively resistant to disinfection, *C. perfringens* spores must be removed by some form of filtration as terminal disinfection is unlikely to inactivate them. Their presence in finished waters, therefore, suggests deficiencies in treatment filtration processes. It has been proposed that the detection of *C. perfringens* spores in finished water may indicate the potential for protozoan cysts to have passed through the treatment process.

International standardised methods are available (ISO 6461-1; 6461-2) and methods for detection of clostridia are relatively easy to perform, even though a simple pasteurisation step is required for the enumeration of spores and strict anaerobic conditions are needed for *Clostridium perfringens*. Clostridia detection only requires a basic routine bacteriology laboratory. They are not normally a health risk for laboratory workers but they are pathogenic and if carelessly handled can give rise to food poisoning and wound infections.


### 2.2.8    *Pseudomonas aeruginosa and Aeromonas spp.*

*Aeromonas* and *Pseudomonas* spp. are Gram-negative, rod-shaped, oxidase positive, non-spore-forming bacteria that are environmentally widespread, with some being opportunistic pathogens.

*Ps. aeruginosa* is commonly found in faeces, soil, water, and sewage but cannot be used as an index of faecal contamination, since it is not invariably present in faeces and sewage, and may also multiply in the enriched aquatic environment and on the surface of organic materials in contact with water.

However, its presence may be one of the factors taken into account in assessing the general cleanliness of water distribution systems. Its presence may lead to deterioration in bacteriological quality, and is often associated with a rise in water temperature or low rates of flow in the distribution system, and consequent complaints about taste, odour, and turbidity.

*Aeromonas* shows no particular association with faecal pollution. Most drinking water treatment processes reduce the numbers of *Aeromonas* to below detectable levels, but treated distributed water can contain larger numbers as a result of regrowth in mains and storage reservoirs. Regrowth of *Aeromonas* depends on the organic content of the water, temperature, the residence time in the distribution network and the presence of residual chlorine (WHO, 2001).

Neither *Pseudomonas* nor *Aeromonas* are indices of faecal pollution, but they may be useful in assessing regrowth in distribution systems. They are both detectable by simple, and inexpensive cultural methods that that can be applied in a basic routine bacteriology laboratory. They may, however, pose a health risk for laboratory workers as some strains of these bacteria are pathogenic. *Ps. aeruginosa* is an opportunistic pathogen that mainly gives rise to superficial infections following contact with heavily contaminated water (but does not cause enteric infections by ingestion). Strains of *Aeromonas* have been implicated in enteric infection but there is no strong evidence that the strains found in water distribution systems are of these types and lead to enteric infection (WHO, 2001). *Aeromonas* strains may also cause wound infections.

### 2.2.9    *Presence-absence test*

Recognising that for good quality drinking waters the majority of samples should not contain any index/indicator organisms, and the detection of any such organism requires action, Clark (1968) developed simple presence-absence tests. Although not strictly speaking a parameter, the presence-absence technique (P-A) can be an economical alternative to coliform analysis. The need to determine the actual number of coliforms within all samples has been questioned, especially in light of the fact that studies have shown that these organisms tend to be irregularly distributed (Pipes and Christian, 1984). The P-A test, which is in essence the most probable number method reduced to a single tube, simply gives an indication of whether coliform bacteria are present or not (Clark, 1980). The test eliminates the types of errors associated with more complex enumeration techniques and record keeping. P-A testing is an effective screening device when assurance of zero coliform organisms is required on a large number of samples. However, it is not an appropriate test where contamination is common and, thus, it is not recommended for use in the analysis of surface water, untreated small-community supplies or larger water supplies that may experience occasional operational and maintenance difficulties. Only a minimal amount of analytical experience is required for the person performing P-A testing because of the simplicity of the methods that are available. Tests have been developed that permit the simultaneous detection of total coliforms and *E. coli*. The P-A test is given as a standard procedure in APHA, AWWA, WEF (1998) and with appropriate confirmatory procedures a single test can detect total coliforms, *Aeromonas*, *Clostridium*, *E. coli*, faecal streptococci, *Pseudomonas* and *Staphylococcus*.

### 2.2.10    Hydrogen sulphide test

Manja *et al.* (1982) developed a very simple screening test for faecal pollution of water sources based on the detection of $H_2S$ production by bacteria. Hydrogen sulphide ($H_2S$) is produced by some bacteria that are associated with faecal contamination, such as some members of *Enterobacteriaceae* (*e.g.* Citrobacters) and some other bacteria (sulphite-reducing clostridia, such as *Clostridium perfringens*). However, a variety of other bacteria not associated with faecal contamination are also capable of producing $H_2S$ under certain conditions. Some bacteria produce $H_2S$ by reducing sulphate and other oxidised forms of sulphur, while other bacteria produce $H_2S$ by degradation of organic sulphur in amino acids and other organic constituents of biomass. The current status, advantages and limitations of $H_2S$ testing for faecal contamination of water was recently reviewed (Sobsey and Pfaender, 2002).

Using a culture medium with thiosulphate as a sulphur source and ferric ammonium citrate as an 'indicator', certain bacteria will produce $H_2S$. The presence of a heavy metal, such as iron salts, in the medium inhibits some bacteria, although *Salmonella*, *Citrobacter* and *Proteus* are all able to produce $H_2S$ in such a medium. The $H_2S$ test uses a treated paper strip that is incubated with the water sample. If bacteria capable of producing $H_2S$ under the test conditions are present in the sample, the production of $H_2S$ turns the paper black. The test can also indicate the severity of contamination if it is used in a semi-quantitative manner by testing dilutions of the sample. Since its initial development, many modifications of the original $H_2S$ test have been reported in the literature and now a number of different $H_2S$ tests are available. Because of the lack of standardisation of these different $H_2S$ tests, there is not a consensus $H_2S$ test method that can be recommended for use. Furthermore, it has not been established that $H_2S$ tests always detect $H_2S$-producing bacteria exclusively associated with faecal contamination. Therefore, it is possible that the test may detect other, non-faecal $H_2S$-producing bacteria from natural sources, leading to a 'false positive' result in terms of faecal contamination.

Despite its limitations, the $H_2S$ strip test is a potentially useful tool for screening water sources and drinking water for faecal pollution, especially for small communities without access to water testing laboratories, or as a simple, initial, warning system. Correlations between the $H_2S$ method and standard faecal pollution microbial indices have been reported even if the test is carried out at room temperature (*i.e.* without incubation). However, $H_2S$ tests are not recommended as substitutes for more specific and better established microbiological parameters for faecal contamination, such as *E.coli*.

### 2.2.11   Other microorganisms

Other microorganisms (*e.g.* bifidobacteria, candida/yeasts, acid-fast bacteria etc.) have been considered in the past as potential parameters of drinking water quality. None of these has been widely accepted and they are not recommended as parameters for routine drinking water evaluation.

### 2.2.12   Pathogens

Various pathogenic microorganisms have been suggested as indices of faecal pollution or indicators of treatment efficiency. However, this approach does not provide the degree of public health protection afforded by the traditional non-pathogenic index or indicator organisms as it depends upon detecting an *actual* risk of infection rather than the *potential* of one. It is also impossible to monitor for all known pathogens and there are also pathogenic agents as yet unrecognised. Nevertheless, pathogen monitoring can provide relevant additional information to that provided by the traditional microbial parameters, particularly in a research context (*e.g.* in validating the significance of indicators of treatment efficiency). Information relating to the presence of pathogens in drinking water is also valuable in the investigation of possible waterborne disease outbreaks (see Chapter 7). Although if monitoring is only instituted when the outbreak has been recognised its value may be much reduced due to the time lag between exposure and development of disease. Transient contamination events or treatment aberrations may have resolved by the time disease occurs and is recognised (Allen *et al.*, 2000).

At present, pathogen monitoring should generally be considered for specific purposes such as research studies, watershed evaluations to target point sources of faecal contamination, outbreak investigations, research into treatment efficiency, and so on. Routine pathogen monitoring based on spot samples has not provided reliable data on their occurrence in source water or treated water. The finding of low numbers of specific pathogens in treated water has, on occasion, resulted in major responses, without any indication of a public health problem. Pathogen detection in treated water should, however, always result in further investigation/evaluation and consideration/assessment of the need for urgent response.

The detection and enumeration of pathogens by culture methods should only be carried out by qualified staff, in specialised laboratories with the proper biosafety equipment and procedures. While most pathogens are present in low numbers in the environment, culturing results in a potential exposure to very high numbers of microorganisms. Molecular, chemical or immunological methods may present less risk, but concentration of large volumes of water still exposes the laboratory worker to a level of risk that requires evaluation and control.

*Enteric viruses*: Discharges of sewage and human excreta constitute the main source of human enteric viruses in the aquatic environment and enteric viruses are always associated with human or animal faecal pollution. However, failure to detect them does not indicate the absence of faecal pollution because their occurrence in faeces is highly variable. They can survive for long periods in the environment and are quite resistant to treatment.

Their enumeration can be expensive and results can take several weeks to obtain if molecular methods are not used (see Chapter 8). Furthermore, many cannot be grown under laboratory conditions. Their detection requires a very well-equipped laboratory and highly trained personnel. In addition, most enteric viruses are pathogenic (to human or animals), albeit at different levels of severity, and virus culture must only be carried out by suitably qualified staff in specialised laboratories with the proper biosafety equipment and procedures.

**Protozoan parasites:** *Cryptosporidium* oocysts and *Giardia* cysts are associated with human and animal faecal sources including amphibians, birds, and mammals, although the species capable of infecting man are restricted to warm-blooded hosts. However, the failure to detect cysts or oocysts does not constitute an indication of the absence of faecal pollution, as their numbers in faeces are highly variable. They can survive for very long periods in the environment and are quite resistant to treatment. They are sometimes found in treated water, usually in low numbers, and when found in filtered supplies suggest deficient coagulation-filtration processes. Viability is difficult to assess but even if non-viable their presence is an indicator of deficient physical treatment and the potential for viable (oo)cysts to be present at some time. Continuous sampling has some value in detecting short-term perturbations in treatment. As with enteric viruses, many species are pathogenic and their isolation and enumeration is expensive and requires a very well equipped laboratory with the proper biosafety equipment and procedures, and highly trained personnel (see also Chapter 2.3.7).

**Table 2.1. Microbial parameter and assay characteristics**

| Parameter | Association with faecal material (*i.e.* pathogens) | Risk to analyst (*i.e.* pathogenicity) | Speed of measurement | Cost | Technical difficulty | Survival in the environment | Resistance to treatment |
|---|---|---|---|---|---|---|---|
| Total coliforms | | L | M | M | M | M | L |
| Thermotolerant coliforms | M | M | M | M | M | M | L |
| *Escherichia coli* | H | M | M | M | M | M | L |
| Faecal streptococci (enterococci) | M | M | M | M | M | M | ISD |
| Ratio between counts (any parameter) | | | | | | | |
| Total bacteria (microscopic) | | | H | M | M | H | H |
| Viable bacteria (microscopic) | | | M | M | M | H | M |
| Total bacteria (automated) | | | H | H | M | H | H |
| Viable bacteria (automated) | | | H | H | M | H | M |
| Heterotrophic bacteria | | L | M | M | M | H | H |
| Aerobic spore-forming bacteria | | L | M | M | M | H | H |
| Somatic coliphages | ISD | M | H | M | M | H | M |
| F specific RNA phages | ISD | M | H | M | M | H | H |
| Bacteroides *phages* | ISD | M | H | M | M | ISD | H |
| Sulphite-reducing clostridia | | L | M | M | M | VH | VH |
| Clostridium perfringens | H | L | M | M | M | VH | VH |
| Pseudomonas, Aeromonas | | M | M | M | M | *VH* | L |
| Enteric viruses | | H | L | H | H | H | H |
| *Giardia* cysts | | H | L | H | H | H | H |
| *Cryptosporidium* oocysts | | H | L | H | H | VH | VH |

Key:
L: low. M: medium. H: high. VH: very high. ISD: insufficient data. ■■■: Not applicable.

Table 2.2. Microbial parameter applicability and suitability

| Parameter | Sanitary survey (catchment) | Source water characterisation | Groundwater characterisation | Treatment efficiency (removal) | Treatment efficiency (disinfection) | Treated water | Distribution system (ingress) | Distribution system (regrowth) | Outbreak investigation |
|---|---|---|---|---|---|---|---|---|---|
| Total coliforms | NR | NR | NR | NR | SA | S | SA* | S | S |
| Thermotolerant coliforms | SA | SA | SA | NR | SA | SA | SA* | S | S |
| *Escherichia coli* | S | S | S | S | S | SA | S* | | S |
| Faecal streptococci (enterococci) | SA | SA | | | | | | | S |
| Ratio between counts (any parameter) | NR | NR | NR | | | | | | |
| Total bacteria (microscopic) | | | | SA | SA | | SA | S | S |
| Viable bacteria (microscopic) | | | | SA | SA | | SA | S | S |
| Heterotrophic bacteria | | | | S | S | NR | S | S | |
| Aerobic spore-forming bacteria | | | | S | S | NR | | | S |
| Somatic coliphages | SA | SA | SA | | SA | | | | S |
| F specific RNA phages | SA | SA | SA | | SA | | | | S |
| Bacteroides *phages* | SA | SA | SA | | SA | | | | S |
| Sulphite-reducing clostridia | NR | NR | NR | | | | | | S |
| Clostridium perfringens | SA | SA | SA | SA | | | | | S |
| Pseudomonas, Aeromonas | | | | | | | | S | |
| Enteric viruses | S | S | S | NR | NR | | | | S |
| *Giardia* cysts, *Cryptosporidium* oocysts | S | S | SA | S | NR | | | | S |

Key:
S: suitable. *In distribution systems without residual disinfection. SA: suitable alternative.
NR: not recommended. ISD: insufficient data. ▉ : not applicable.

## 2.3 Non-microbial parameters

In addition to microbial measurements there are also various physicochemical assays of water quality that can provide useful information about the quality of, and changes in the quality of, raw water and the effectiveness of applied treatment processes. Many of the parameters can be analysed relatively quickly and at less cost than the microbial parameters and a number can be measured on-line and can be automated providing real-time data that can be linked to alarms or process control equipment. Non-microbial parameters are outlined below and summarised, at the end of the section, in Tables 2.3 and 2.4.

For most non-microbial assays, the benefit of their use comes from the speed and simplicity of measurement rather than the specificity of the assay itself. The value of the tests comes from their application as triggers, to give early warning through the detection of changes or unusual events, which can then be followed up more rigorously.

### 2.3.1 *Rainfall events*

Rainfall events are one of the most important causes of degradation in source water quality affecting surface waters and ground waters (see Chapter 4). Rainfall drives the movement of pathogens into and through water bodies and can move soil, resuspend sediments, cause overflow of combined and poorly maintained sewers, degrade groundwater through infiltration and so on. Forecasting and rainfall detection systems such as radar, hydrographic monitoring equipment and remote sensing can now be used to provide authorities with advanced warnings of upcoming rainfall events that might influence water quality and treatment. Stream flows and heights can be measured on-line or manually to warn of major hydrological events. Although not a measure of faecal loading, rainfall events are useful in predicting deterioration in source water quality and permit appropriate precautionary measures to be taken to safeguard treated water quality.

### 2.3.2 *Flow*

Measurement of flow of surface waters as well as flow during drinking water treatment provides important information regarding the availability and production of quality water. Low flow in surface waters may lead to biological degradation and higher concentrations of pollutants due to reduced dilution of discharges. During treatment, changes in flow can adversely affect coagulation and sedimentation processes, while filtration rate and contact time with disinfectant are key factors in the production of safe drinking water. Flow is easily measured using continuous on-line measurements. Changes in flow rates within distribution systems can result in suspension of sediments and deterioration of supplies.

### 2.3.3    Colour

Colour in drinking water may be due to the presence of coloured organic matter, *e.g.* humic substances, metals such as iron or manganese, or highly coloured industrial wastes. The appearance of colour in water is caused by the absorption of certain wavelengths of light by coloured substances ('true' colour) and by the scattering of light by suspended particles, together these are termed 'apparent' colour. Treatment removes much of the suspended matter and, generally speaking, drinking water should be colourless. Source waters high in true colour can be treated to remove colour by oxidation with ozone and adsorption onto activated carbon.

Changes in colour from that normally seen can provide warning of possible quality changes or maintenance issues and should be investigated. They may, for example, reflect degradation of the source water, corrosion problems in distribution systems, changes in performance of adsorptive treatment processes (such as activated carbon filtration) and so on. It is simply and cheaply measured using a spectrophotometer or simple colorimeter or using visual comparison with known standards.

### 2.3.4    pH

The pH of water affects treatment processes, especially coagulation and disinfection with chlorine-based chemicals. Changes in the pH of source water should be investigated as it is a relatively stable parameter over the short term and any unusual change may reflect a major event. pH is commonly adjusted as part of the treatment process and is continuously monitored.

Equipment for continuous monitoring and data logging is available at reasonable cost. Simple paper strip or colorimetric tests are also available, while these are less precise they can provide valuable information. Test methods are inexpensive, require average skill, and are performed routinely by many laboratories and may easily be conducted on site.

### 2.3.5    Solids

Water always contains a certain amount of particulate matter ranging from colloidal organic or inorganic matter that never settles to silts, algae, plankton or debris of all kinds that can settle quite rapidly. Various methods have been devised to identify or measure these solids. In raw water storage reservoirs and other large bodies of water, discs can be used to measure the depth of water through which the disc remains visible (*i.e.* transparency). Suspended solids can be measured indirectly as turbidity, which depends on the scattering of light by particles in water (see 2.3.6) or by particle size-counters (see 2.3.7). Nephelometers, particle size analysers and physico-chemical methods provide more precise measurements of the solids in water.

Solids can be dissolved solids or suspended solids in water: together they are referred to as total solids. They can be measured directly, separately or together by physico-chemical methods using combinations of filtration and evaporation. What is left after evaporation of the water before and after filtration through a 2.0 µm filter are referred to respectively as 'total solids' and 'total dissolved solids'. Material retained on the filter is referred to as 'total suspended solids'. Methods for the measurement of solids are well described (APHA, AWWA, WEF, 1998) and involve simple procedures such as filtration, evaporation and/or drying at specified temperatures, and weighing. Results are reported in mg/l.

The amount of solids in water affects both removal and disinfection processes. The solids content of waters can vary significantly with seasons and rainfall events. Abnormal changes in the amount or type of solids in source or treated water should be investigated. Solids, whether total or dissolved, can provide information on the pollution level of the water. Solids can affect taste and appearance of drinking water. Furthermore, a significant increase in the levels of solids could be related to contamination from a range of sources such as freshly derived surface run-off, ingress or wastewater.

Conductivity assays can be used to reflect total dissolved solids concentrations and can be applied rapidly on-line, although conductivity mainly reflects the mineral content. In relatively low salinity waters a marked change in conductivity can provide an indication of contamination with more saline waters, such as most types of wastewater (as wastewaters are typically more than an order of magnitude more saline than surface freshwater).

Many of the tests methods for solids are inexpensive, some can be undertaken in-field or on-line, most require average skill, and others can be performed routinely by many laboratories providing data within hours.


### 2.3.6    *Turbidity*

Turbidity is a measure of suspended solids. It has been singled out here because it is probably the most generally applicable and widely used non-microbial parameter that can provide the most significant data throughout the water abstraction and treatment process. It is not associated specifically with faecal material, but increases in turbidity are often accompanied with increases in pathogen numbers, including cysts or oocysts. Turbidity is often determined by measuring the amount of light scattered by the particulate matter in the water using a nephelometer. Instruments for measuring turbidity are calibrated using commercially available certified standardised suspensions of formazin defined in Nephelometric Turbidity Units (NTU). The lowest level measurable by modern nephelometers is about 0.02 NTU. Nephelometers are available as on-line continuous turbidity meters and they can provide precise data on variations in water treatment efficiency. Data can be collected electronically and stored in a digital format for reporting, analysis or as part of a process-control scheme.

Waterworks using filtration should be able to achieve values of 0.5 NTU or less. Regulations in various countries specify values from 0.1 to 5 NTU in final treated water. Where financial resources are not available for continuous monitoring, manual measurements at regular and frequent intervals can be obtained using simple portable low cost instruments. Some of these are simple comparator devices. A very cheap turbidity measurement method is based on transparency, which can be used to measure down to 5 NTU, this is useful in terms of small community supplies where a high level of sensitivity is not necessary (WHO, 1997).

The turbidity of water affects treatment processes and especially disinfection with chlorine-based chemicals. It is important to know the turbidity characteristics of water sources and to respond to unexplained changes in turbidity. Turbidity of surface water sources may be heavily influenced by rainfall events or algal growth and treatment processes should be tailored to respond to such changes. Most groundwaters have a relatively stable turbidity and any change reflects a major event that needs to be investigated and corrected by tailoring the treatment to the incoming water quality. Even relatively small changes may be important and outbreaks of cryptosporidiosis have been associated with small changes in turbidity of relatively short duration (Waite, 1997). Turbidity is also a good measure of the extent to which treatment processes remove suspended matter. Turbidity of filtered water should be monitored at each filter and data above the expected values should be investigated. Monitoring of the combined filtrate alone from a number of filters may not detect significant loss of performance of an individual filter. This is particularly important in relation to the removal of cryptosporidial oocysts as they are not inactivated by conventional disinfection, and effective filtration is the only treatment means for their control.

Equipment for continuous monitoring and data logging is available at relatively low cost. Test methods are inexpensive, require average skill, and are performed routinely by most laboratories.

### 2.3.7    *Particle size analysis*

Particles in water are distributed over a wide range of size. Enumeration and identification of the nature of the larger particles is best achieved by microscopy (see section 2.3.8). Various other instruments have been developed to enumerate and size particles in water. These instruments measure the passage of particles in a sensing zone where each is counted and sized according to the electronic pulse generated. This pulse is proportional to the characteristics of the size and shape of the particle. The apparatus generates a report on the number of particles in each size-class selected. There are different types of instruments available, but they are all computerised, often complex and expensive. They also require careful calibration in order to generate data that is comparable between different instruments. They are especially useful in determining filtration efficiency during drinking water treatments. The surveillance of the removal of particles in the 2-5 micrometer size range (*i.e.* the size of cysts *Giardia* or oocysts of *Cryptosporidium*) is currently being evaluated as a potential surrogate for their removal.

Particle counting can provide a general index of removal effectiveness and as such is a good quality control parameter for filtration. However, factors other than size (such as electric charge on the particles) may affect removal processes. Particle size monitors are available as on-line instruments, however as mentioned earlier, the equipment is expensive and it requires a greater level of skill than turbidity analysis.

### 2.3.8    *Microscopic particulate analysis*

Microscopic particulate analysis provides detailed microscopic information on the nature of particulates in water. Biological particles (cysts, diatoms, fungi, zoo-plankton and phyto-plankton) and inorganic particles are described and enumerated. It is useful to identify contaminants in groundwater, providing information on the nature and likely origins of its contamination. Groundwater influenced by surface water will contain a significant amount of algae and other particles not normally found in protected groundwater. It is mainly of value as a research and investigational tool (see Chapter 7) rather than for routine monitoring. The analysis requires well-trained skilled personnel, is time-consuming and is performed by few laboratories.

### 2.3.9    Disinfectant residual concentration

Chlorine is the most widely used disinfectant in water treatment. For the majority of bacterial pathogens, and some viruses, terminal disinfection is the critical control point of treatment and proper measurement and control of disinfectant dose and contact time (alongside pH and turbidity) is imperative. The measurements of disinfectant dose, residual obtained and the time of contact are primary data that provide a minimal level of quality control of treated water and disinfectant residual concentration during and after disinfection is a required measurement at most water treatment works. Wherever possible residual concentration after contact should be continuously monitored, with suitable alarms to signal departures from the pre-set target range and, in some cases, provision for automatic shutdown of the treatment process may be appropriate. Instruments for continuous monitoring and data logging are also available at reasonable cost. Simple and inexpensive colorimetric tests using titration methods or kits are available for manual determination by relatively low skilled personnel.

### 2.3.10    Organic matter

Data on the level of organic matter in treated water provide an indication of the potential for the regrowth of heterotrophic bacteria (including pseudomonads and aeromonads) in reservoirs and distribution systems. Organic matter can be measured as Total Organic Carbon (TOC), Biochemical Oxygen Demand (BOD) or Chemical Oxygen Demand (COD). BOD is primarily used with wastewaters and polluted surface waters, and TOC is the only parameter applicable to drinking water. Measurement of these three parameters requires basic laboratory facilities and adequately trained personnel. TOC measurement can now be obtained using on-line instrumentation. The data provide information on the amount of matter present in the water. Not all organic matter is biologically available and it may be useful to measure the amount of organic material available to support bacteriological growth. Although BOD does this to a degree, a number of other measurements such as assimilable organic carbon (AOC) have been proposed. These latter methods require skilled personnel and a well-equipped laboratory.

### 2.3.11    Specific chemical parameters

Ammonia is rapidly oxidised in the environment and is typically found in natural waters at concentrations less than 0.1 mg/l. Concentrations significantly above this indicate gross contamination by fresh sanitary waste, where ammonia levels are typically very high (tens to hundreds of mg/l). Relatively simple and rapid in-fields tests are available for ammonia that could be used as an initial screen. Ammonia combines readily with chlorine to produce chloramines, which are much less effective disinfectants but are more stable.

Boron measurement has been proposed as an index of faecal pollution. Salts of boron have been used as a water softener, and calcofluor as a whitener in detergents, resulting in their presence in wastewater. Their use, however, is widely being discontinued, which markedly reduces the value of these parameters as indices of sewage/wastewater pollution.

Excreted materials, such as faecal sterols, secretory immunoglobulin type A, caffeine, urobilin and a number of other parameters have been suggested as indices of faecal pollution. Their detection and measurement usually require well-equipped laboratories with trained personnel. Research on the use of these parameters is ongoing and, as yet, they are not recommended for routine monitoring.

**Table 2.3. Non-microbial parameter assay characteristics**

| Parameter | Speed of measurement | Possibility of online monitoring or automation | Cost | Technical difficulty |
|---|---|---|---|---|
| Rainfall events | H | H | L | L |
| Flow | H | H | L | L |
| Colour | H | H | L | L |
| pH | H | H | L | L |
| Solids (Total and dissolved) | M | *L* | M | M |
| Conductivity | H | H | L | L |
| Turbidity | H | H | L | L |
| Particle size analysis | H | H | H | H |
| Microscopic particulate analysis | H | *L* | H | H |
| Disinfectant residual | H | H | L | L |
| Organic matter (TOC, BOD, COD) | M | H | M | M |
| Ammonia | H | M | M | M |
| Detergents (Boron, calcofluor) | | | | |
| *Faecal sterols, IgA (secretory), caffeine, urobilin* | | | | |

**Key**: L: low. M: medium. H: high. ▓▓▓ : not applicable.

71

**Table 2.4. Non-microbial parameter applicability and suitability**

| Parameter | Sanitary survey (catchment) | Source water characterisation | Groundwater characterisation | Treatment efficiency (removal) | Treatment efficiency (disinfection) | Treated water | Distribution system (ingress) | Distribution system (regrowth) | Outbreak investigation |
|---|---|---|---|---|---|---|---|---|---|
| Rainfall events | S | | | | | | | | S |
| Flow | S | S | | | S | S | | S | S |
| Colour | | | | | | S | | | S |
| pH | | | | | S | | | | S |
| Solids (Total and dissolved) | S | S | S | | | | | | S |
| Conductivity | S | S | S | | | | | | S |
| Turbidity | S | S | S | S | | | | | S |
| Particle size analysis | | | | S | | | | | S |
| Microscopic particulate analysis | | | S | | | | | | S |
| Disinfectant residual | | | | | | S | S | SA | S |
| Organic matter (TOC, BOD, COD) | S | S | S | | | | | S | S |
| Ammonia | S | S | S | | | | | ISD | S |
| Detergents (Boron, calcofluor) | NR | NR | NR | | | | | | |
| *Faecal sterols, IgA (secretory), caffeine, urobilin* | | | | | | | | | |

Key:
S: suitable. SA: suitable alternative. ISD: insufficient data. NR: not recommended.
█ : not applicable.

## 2.4    Summary

For drinking water to be wholesome it should not present a risk of infection, or contain unacceptable concentrations of chemicals hazardous to health and should be aesthetically acceptable to the consumer. The infectious risks associated with drinking water are primarily those posed by faecal pollution, and their control depends on being able to assess the risks from any water source and to apply suitable treatment to eliminate the identified risks. Rather than trying to detect the presence of pathogens, at which time the consumer is being exposed to possible infection, it is practice to look for organisms, while not pathogens themselves, that show the presence of faecal pollution and therefore the potential for the presence of pathogens. A number of microbial parameters have been used as 'index' organisms to give an indication of the amount of faecal pollution of source waters, the pre-eminent being *E. coli*. It is also important to be able to check on the effectiveness of treatment processes at eliminating any pathogens that might have been present in the untreated source, and 'indicator' organisms fulfil that role. While the perfect indicator needs to be as resistant to treatment processes as the most resistant potential pathogen, no single parameter is ideal. In principle, treatment should be able to eliminate all non-sporing bacteria and enteric viruses and the less restricted the parameter chosen the more suitable it should be. There are a number of microbial parameters that are of some value as indices or indicators and these are discussed. Water quality can deteriorate in distribution due to ingress or regrowth and measures of regrowth potential are described.

A number of non-microbial parameters are described, which can provide useful information about quality, and changes in quality, of source waters and the effectiveness of treatment processes.

# REFERENCES AND FURTHER READING

Allen, M.J., Clancy, J.L. and Rice, E.W. (2000) Pathogen monitoring – old baggage from the last millennium. *Journal of the American Water Works Association* **92**(9), 64-76.

Anon (1994) *The Microbiology of Water 1994: Part 1 – Drinking water.* Reports on Public Health and Medical Subjects, No. 71. Her Majesty's Stationery Office, London.

Anon (1999) *Waterborne pathogens.* AWWA Manual of water practices, M48. American Water Works Association, Denver, Colorado.

Armon, R. and Kott, Y. (1995) Distribution comparison between coliphages and phages of anaerobic bacteria (*Bacteriodes fragilis*) in water sources and their reliability as faecal pollution indicators in drinking water. *Water Science and Technology* **31**(5-6), 215-222.

APHA, AWWA, WEF (1998) *Standard Methods for the Examination of Water and Wastewaters, 20th Edition.* American Public Health Association, Washington, DC.

Bartram, J. and Ballance, R. eds. (1996)*Water Quality Monitoring. A practical guide to the design and implementation of freshwater quality studies and monitoring programmes.* E & FN Spon, published on behalf of World Health Organization and the United Nations Environment Programme

Chin, J. (2000) *Control of Communicable Diseases Manual. 17th Edition.* American Public Health Association, Washington, D.C.

Clark, J.A. (1968) A presence-absence (PA) test providing sensitive and inexpensive detection of coliforms, fecal coliforms and fecal streptococci in municipal drinking water supplies. *Canadian Journal of Microbiology* **14**, 13-18.

Clark, J.A. (1980) The influence of increasing numbers of non indicator organisms on the membrane filtration and P-A tests. *Canadian Journal of Microbiology* **15**, 827-832.

Fujioka, R., Sian-Denton, C., Borja, M., Castro, J. and Morphew, K. (1999) Soil: the environmental source of *Escherichia coli* and enterococci in Guam's streams. *Journal of Applied Microbiology Symposium Supplement* **85**, 83S-89S.

Gleeson, C. and Gray, N. (1997) *The Coliform Index and Waterborne Disease*. E and FN Spon, London. pp194.

Grabow, W.O.K. (2001) Bacteriophages: Update on application as models for viruses in water. *Water SA* **27**(2), 251-268.

Hsu, F.-C., Shieh, Y.-S.C. and Sobsey, M.D. (1995) Genotyping male-specific RNA coliphages by hybridization with oligonucleotide probes. *Applied and Environmental Microbiology* **61**, 3960-3966.

Hurst, C.J., Knudsen, G.R., McInerney, M.J., Stetzenbach, L.D. and Walter, M.V. (2001) *Manual of Environmental Microbiology 2^{nd} Edition*. American Society for Microbiology Press, Washington, DC.

ISO 6461-1 (1986) Water quality – Detection and enumeration of the spores of sulphite-reducing anaerobes (clostridia) – Part 1: Method by enrichment in a liquid medium. International Organization for Standardization, Geneva, Switzerland.

ISO 6461-2 (1986) Water quality – Detection and enumeration of the spores of sulphite-reducing anaerobes (clostridia) – Part 2: Method by membrane filtration. International Organization for Standardization, Geneva, Switzerland.

ISO 9308-1 (1990) Water Quality – Detection and enumeration of coliform organisms, thermotolerant coliforms and presumptive *Escherichia coli* – Part 1: Membrane filtration method. International Organization for Standardization, Geneva.

ISO 9308-2 (1990) Water Quality – Detection and enumeration of coliform organisms, thermotolerant coliforms and presumptive *Escherichia coli* – Part 2: Multiple tube (most probable number) method. International Organization for Standardization, Geneva.

ISO 10705-1 (1995) Water quality – Detection and enumeration of bacteriophages – Part 1: Enumeration of F-specific RNA bacteriophages. International Organization for Standardization, Geneva, Switzerland.

ISO 10705-2 (1995) Water quality – Detection and enumeration of bacteriophages – Part 2: Enumeration of somatic coliphages. International Organization for Standardization, Geneva, Switzerland.

ISO 10705-4 (1995) Water quality – Detection and enumeration of bacteriophages – Part 4: Enumeration of bacteriophages infecting *Bacteriodes fragilis*. International Organization for Standardization, Geneva, Switzerland.

Leclerc, H., Edberg, S., Pierzo, V. and Delattre, J.M. (2000) Bacteriophages as indicators of enteric viruses and public health risk in groundwaters. *Journal of Applied Microbiology* **88**(1), 5-21.

Manja, K.S., Maurya, M.S. and Rao, K.M. (1982) A simple field test for the detection of faecal pollution in drinking water. *Bulletin of the World Health Organization* **60**, 797-801.

Murray, P.R. (1999) *Manual of Clinical Microbiology*. American Society for Microbiology Press, Washington, DC.

Pipes, W.O. and Christian, R.R. (1984) Estimating mean coliform densities of water distribution systems. *Journal of the American Water Works Association* **76**, 60-64.

Sobsey, M.D., Battigelli, D.A., Handzel, T.R. and Schwab, K.J. (1995) *Male-specific Coliphages as Indicators of Viral Contamination of Drinking Water*. American Water Works Association Research Foundation, Denver, Co. pp. 150.

Sobsey, M.D. and Pfaender, F.K. (2002) *Evaluation of the H₂S Method for Detection of Faecal Contamination of Drinking water*. WHO/SDE/WSH 02.08. World Health Organization, Geneva.

USEPA (2001) *Protocol for Developing Pathogen TMDL. 1st Edition*. EPA 841-R-00-002. US Environmental Protection Agency, Office of Water, Washington, DC.

Waite, W.M. (1991) Drinking water standards – a personal perspective. In: *Proceedings of the UK Symposium on Health-related Water*

*Microbiology.*, Morris, R. *et al.* (Eds.). International Association for Water Pollution Research and Control, London. pp. 52-65.

Waite, W.M. (1997) *Assessment of Water Supply and Associated matters in Relation to the Incidence of Cryptosporidiosis in Torbay in August and September 1995.* Drinking Water Inspectorate, London.

WHO (1993) *Guidelines for Drinking water Quality, Volume 1: Recommendations.* World Health Organization, Geneva.

WHO (1996) *Guidelines for Drinking water Quality, Volume 2: Health criteria and other supporting information.* World Health Organization, Geneva.

WHO (1997) *Guidelines for Drinking water Quality, Volume 3: Surveillance and control of community supplies.* World Health Organization, Geneva.

WHO (2001) *Guidelines for Drinking water Quality. Addendum: Microbiological agents in drinking water.* World Health Organization, Geneva.

*Chapter 3*

## ASSESSMENT OF RISK

*P.R. Hunter, P. Payment, N. Ashbolt and J. Bartram*

## 3.1    Introduction

This chapter is primarily about the role of analytical techniques in the assessment of risk and specifically the value of water quality indicator parameters in this process. Assessment of risk in relation to drinking water supplies is undertaken for a number of reasons (Percival *et al.*, 2000):

- To predict the burden of waterborne disease in the community, under outbreak and non-outbreak conditions. This is helpful in determining the impact of improvements in water supply on health and to act as a driver towards improvement.

- To help set microbial standards for drinking water supplies that will give tolerable levels of illness within the populations drinking that water.

- To identify the most cost-effective option to reduce microbial health risks to drinking water consumers.

- To help determine the optimum treatment of water to balance microbial risks against chemical risks from disinfection by-products.

- To provide a conceptual framework to help individuals and organisations understand the nature and risk to, and from, their water and how those risks can be minimised.

The focus of this chapter is to review the value of indicator parameters of water quality and other analyses in the context of three different approaches to the assessment of risk, namely:

- Epidemiological methods.

- Quantitative microbial risk assessment (QMRA).

- Qualitative risk assessment (including risk ranking).

## 3.2    What is risk?

Risk can be defined in the simplest form as 'the possibility of loss, harm or injury'. This definition includes two separate concepts; the probability of an event and the severity of that event. These two concepts are illustrated in Figure 3.1, and this model helps the prioritisation of risks for any risk-reduction action. Clearly those risks that need most urgent action are high probability – high severity risks (upper right quadrant). Those that need little, if any, attention are low probability – low severity (lower left quadrant).

**Figure 3.1. Two-dimensional classification of risk**

| | Low probability of severe harm<br><br>(should be given intermediate priority attention) | High probability of severe harm<br><br>(needs most urgent attention) |
|---|---|---|
| | Low probability of mild harm<br><br>(can probably be ignored or given low priority attention) | High probability of mild harm<br><br>(should be given intermediate priority attention) |

*Severity of harm* (left axis)

**Probability of occurrence**

Despite the simplicity of this two-dimensional model, the processes that allow the calculation or quantification of risk differ. Indeed, many risk-based decisions are still subjective or semi-quantitative. Even where risk assessments are presented in an apparently objective, numerical manner these are often based on assumptions which are themselves subjective or semi-quantitative. One of the major problems with all forms of assessing risk is the quality and levels of uncertainty in much of the basic data (Macgill *et al.*, 2001).

## 3.3    Types of evidence

Data used in the assessment of risk is obtained from experimental work on animals or volunteers and from epidemiological investigations. These epidemiological investigations may be conducted during an outbreak investigation or be done as part of planned research to investigate the causes and transmission of disease.

The most abundant source of epidemiological data on waterborne disease comes from outbreak investigations (Chapter 7), and outbreaks provide very valuable data for the assessment of risk. Particularly, outbreaks can provide clear evidence that a specific pathogen can be spread by the water route. Outbreak investigations also provide good information on what failures in the water supply and distribution chain led to the risk to health. This enables risk management strategies to focus on those stages in the water supply chain where failures are likely to occur. Outbreaks can also be the setting for epidemiological studies that provide useful information on what non-water-related factors affect risk of infection with the outbreak pathogen. However, outbreak data have their limitations (Andersson and Bohan, 2001). For any particular pathogen, it is rarely known what proportion of the burden of disease is due to sporadic spread by the water route. Nor is it known whether those factors responsible for failure leading to outbreaks are also those factors responsible for sporadic disease. Consequently information reliant only on outbreaks may not be applicable to the major proportion of waterborne disease. Also, epidemiological investigations of water-related disease may be biased by prior knowledge of cases and controls about the suspected cause of the outbreak (Hunter, 2000).

Targeted epidemiological studies can provide good data on the relationship between specific water quality parameters and disease in a population. Such studies can identify relationships between risk factors for all waterborne disease and not only that associated with outbreaks. Separating the waterborne fraction of gastrointestinal disease from the numerous other routes of infection is a challenge and the results from most epidemiological studies are presented as a level of association between drinking water and the parameter(s) under study. These studies are often subject to criticism as there are rarely clear-cut conclusions, and they are potentially subject to a number of biases and confounding factors.

Quantitative microbial risk assessment (QMRA) is an emerging field that has applications in specific situations and is discussed in more detail below. QMRA uses information on the distribution and concentration of specific pathogens in the water supply along with information on the infectivity of those pathogens to determine risk to public health.

Assessment of the quality of evidence is important yet rarely formally addressed in the assessment of risk (Macgill *et al.*, 2001). Requirements for evidence related to demonstration of causality may be very different to that for dose response. In practice the overall body of evidence may include a number of studies each with strengths and weaknesses and employing often very different methods and approaches (Blumenthal *et al.*, 2001; Haas and Eisenberg, 2001).

When assessing the risk of disease due to drinking water it is very important to consider the overall body of evidence, weighing each piece of evidence as to its quality. Given the uncertainty inherent in all epidemiological studies reliance on a single study, even an extremely well conducted one, may be misleading.

## 3.4 Epidemiological approaches to risk

Epidemiology is the study of the incidence and transmission of disease in populations. Epidemiological investigations are central to the assessment of risk (Blumenthal *et al.*, 2001), both in providing estimates of risk and in providing input data into risk assessment models. The epidemiological definitions of risk are distinct from definitions used more generally, and are defined in Table 3.1.

**Table 3.1. Epidemiological definitions of risk**

| Risk | Definition |
|---|---|
| Absolute risk | The number of new cases occurring within a certain sized population during a specified time period, usually referred to as incidence. |
| Attributable risk | The proportion of cases of a disease due to a particular risk factor. |
| Relative risk | The ratio between the incidence of disease in those members of the population exposed to a possible risk factor and those not exposed. |
| Odds ratio | The ratio between the probability that someone with a disease has experience of the potential environmental factor and the probability that a control has experience of the same factor. Provides an estimate of relative risk in case control studies. |

Epidemiology relies on a limited range of methods and approaches to define risk (discussed in more detail elsewhere, *e.g.* Gordis, 2000). Most epidemiological studies can be classified as descriptive, analytical or intervention. Descriptive epidemiological studies set out to describe the distribution of cases of disease in time, place and person. Two types of descriptive study that have been used in relation to waterborne disease are the ecological study and the time series study. Analytical studies are generally of the case control or cohort type, in which individuals or groups are compared. Intervention studies are experimental studies that observe the impact of certain interventions (such as provision of point-of-use filters) on the risk of illness. The various types of study are described in Table 3.2.

**Table 3.2. Types of epidemiological study that have been used in risk assessment of waterborne disease**

| Study type | Description | Advantages and disadvantages |
|---|---|---|
| Ecological study | Determining relationship between disease and risk factors by comparing the incidence of disease in different communities with varying exposure to risk factors. | Relatively inexpensive to carry out providing that disease rates and data on risk factors are already available. Because data is only available for groups, it is not known whether individuals with disease are exposed to risk factor. Good for generating hypotheses, but cannot be used as evidence of epidemiological proof. |
| Time series study | Determining relationship between disease incidence in a population and variation in a risk factor over time. | A type of ecological study and subject to the same advantages and disadvantages. |
| Case-control study | Determining relationship between disease and risk factors by comparing the incidence of disease in exposed individuals to matched controls. | Relatively inexpensive to carry out. Generates data on individuals exposed to the risk factors in comparison with healthy individuals. |
| Cohort study | Comparing rate of disease in two, or more, populations with different levels of exposure over a specific period of time on randomly selected individuals. | Relatively expensive to carry out. Generates data on the risk factors in populations by comparing groups of randomly selected individuals. |
| Intervention study | Comparing the rates of disease in two or more groups (cohorts) of randomly chosen individuals after intervening to change the level of exposure. | The gold standard for epidemiological proof, but can be time consuming and costly to carry out. |

## 3.5    Studies linking ill health to indicators

While many microorganisms have been implicated as causative agents in outbreaks of various diseases, there is little epidemiological data on the endemic level of waterborne diseases and their aetiology. The association between many aetiological gents with a given route of exposure and their contribution to the total burden of disease is often uncertain. Studies that have attempted to define the burden of waterborne disease have targeted gastrointestinal illness, as it is the most frequent and easy to measure adverse outcome associated with drinking water (Prüss *et al.*, 2002). This frequent outcome enables researchers to obtain information faster than with less common outcomes (*e.g.* hepatitis) or outcomes that are less defined and are more difficult to link with specific exposures (*e.g.* malignant disease). However, use of gastrointestinal disease as an index of water-related disease impact has a number of limitations. Depending on how gastroenteritis is measured estimates of disease burden can vary substantially. Since the disease may be considered 'mild', especially amongst adults, relatively few people seek medical attention and even if they do they may not have faecal samples taken for laboratory investigation. Consequently, disease burden estimates based on national surveillance systems of laboratory reports can substantially underestimate disease burden (Wheeler *et al.*, 1999). This has led to the use of self-reported gastroenteritis in several studies (discussed below). There are, however, problems with the use of self-reported gastroenteritis as a marker of disease, as depending on how gastroenteritis is defined rates can vary substantially. How the data is collected can also markedly affect estimates of disease burden. Retrospective studies, where individuals are asked whether they have had diarrhoea in the previous month can over-estimate illness by about three times when compared to prospective studies where volunteers maintain a daily health diary (Wheeler *et al.*, 1999). This overestimate may be greater in outbreak settings (Hunter and Syed, 2001). Furthermore, since gastrointestinal disease is relatively common and may be transmitted by various means, it may be difficult to distinguish the waterborne contribution from the background 'noise'.

The link between substandard drinking water and disease is relatively easy to demonstrate. Such a demonstration becomes more difficult to make as the quality of the water improves towards the current World Health Organization (WHO) Guidelines (WHO, 1993; 1996; 1997). Indeed, the link between highly treated drinking water meeting local regulations, as found in most industrialised countries, and microbial illness has only been reported relatively recently. For example, both waterborne *Giardia* and *Cryptosporidium* infection have clearly been linked to drinking water meeting or exceeding current standards, thereby challenging the value of the traditional microbial indicator parameters as well as the efficacy of treatment procedures (Gostin *et al.*, 2000).

### 3.5.1    Untreated drinking waters

In developing countries there is abundant evidence that poor quality water containing indices of faecal pollution is the source of much disease in the population. There is, however, little data on the exact relationship between the two.

There is a substantial body of evidence that relates improvements in water supply and sanitation in general and in drinking water quality in particular, to specific health outcomes (most frequently reductions in diarrhoeal disease). Many of the early studies had severe methodological flaws (Blum and Feachem, 1983), but two reviews of published studies have sought to identify better-conducted studies and assess the detected disease outcomes (Esrey *et al.*, 1985; 1991). Most studies detected were from less-industrialized countries and a median reduction in diarrhoeal disease of 26 − 27% was reported. However, water quality was typically not assessed and in some cases opportunities for recontamination may have cast doubt on the actual intervention tested. In some more recent studies, far better characterisation of the intervention has been achieved with actual water quality measurements made (*e.g.* Quick *et al.*, 1999 [*E. coli*] and Semenza *et al.*, 1998 [chlorine residual]). Nevertheless, the absence of an estimate of exposure from most studies renders them unusable in formalised risk assessment requiring description of population dose-response.

### 3.5.2 Substandard drinking water

In France, Collin *et al.* (1981) prospectively studied gastrointestinal illnesses associated with the consumption of tap water, using reports from physicians, pharmacists and teachers. They reported five epidemics associated with poor quality water but they did not address the endemic level of gastrointestinal illnesses. The same group found a relationship between faecal streptococci and acute gastrointestinal disease (Ferley *et al.*, 1986; Zmirou *et al.*, 1987) in a study of 64 villages with sub-standard water. Thermotolerant coliforms, total coliforms and total bacteria made no independent contribution to disease. Zmirou *et al.* (1995) investigated the effect of chlorination alone, on water that did not satisfy microbiological criteria otherwise. The crude incidence of diarrhoea was 1.4 times more frequent in children from villages where water supplies had evidence of faecal pollution, even after chlorination. In Israel, Fattal *et al.* (1988) addressed the health effects of drinking water and did not show a relationship between health effects and total or thermotolerant (faecal) coliforms. Beaudeau *et al.* (1999) reported a relationship between the chlorine disinfection level and diarrhoeal illness in the population of Le Havre (France).

### 3.5.3 Drinking water meeting current regulations

In the USA, Batik *et al.* (1979) attempted to use cases of hepatitis A as an indicator of health risk, but could not establish a correlation with water quality nor, in a later study, did they find a correlation between traditional indicator parameters (coliforms) and the risk of waterborne outbreaks (Batik *et al.*, 1983).

Craun *et al.* (1997) in the USA, evaluated the relationship between coliform compliance and outbreak occurrence. They found that coliforms were usually found in the water during an outbreak investigation but that during the previous months, coliforms were detected in only half of the systems and caused a violation in only a quarter of them. The violation rate was not different between community systems that experienced an outbreak and those that did not. In Namibia, Isaäckson and Sayed (1988) conducted a similar study and did not observe an increased risk of gastrointestinal illness associated with the consumption of recycled wastewater.

In Canada, two prospective studies have suggested that a very high proportion of gastrointestinal illnesses could still be attributable to tap water consumption, even when water (albeit from a degraded catchment) exceeded the current drinking water quality guidelines (Payment *et al.*, 1991; 1997).

Turbidity of treated drinking water has been linked to health effects in Milwaukee (MacKenzie *et al.*, 1994; Morris *et al.*, 1996), in Philadelphia (Schwartz *et al.*, 1997; 2000) and in Le Havre (Beaudeau *et al.*, 1999). It should be noted, however, that these studies of turbidity and adverse health outcome are 'ecological', in that they measure exposure of populations rather than of individuals and, as such, potentially suffer from bias due to the so called 'ecological fallacy' (Walter, 1991). While this does not mean that these studies are invalid, they cannot be taken as proof of an association in their own right.

### 3.5.4    The role of index/indictor parameters in assessing risk to health

During the course of the 20th century, the absence of traditional index/indicator parameters in drinking water was related to a significant reduction in waterborne outbreaks. This reflected the use of these organisms to indicate the presence of faecal contamination and through which valuable information on effectiveness and failure of interventions was progressively accumulated. More recently, occasional outbreaks and endemic disease have been linked to waterborne disease in the absence of the traditional indicator parameters. The causes are often failures in treatment or contamination of the treated product, but the coliform parameters (total, thermotolerant or *E. coli*) cannot provide information on the removal and inactivation of pathogens that are several orders of magnitude more resistant to treatment. Hence, coliform parameters remain useful for specific purposes described elsewhere in this book, but future studies on waterborne disease should be targeted to additional indicator parameters (for instance, those described in Chapter 2). There is, however, no single direct measurement (including direct pathogen testing) available to predict health outcomes in a population. Turbidity and faecal streptococci counts are the main indicator parameters that have been shown to have independent association with actual levels of disease in populations.

## 3.6 Quantitative microbial risk assessment (qmra)

The QMRA approach to risk differs from epidemiological approaches in that the latter seeks to measure actual levels of disease in the population while the former attempts to calculate risk from what is known, or can be inferred, about the concentration of particular pathogens in the water supply and the infectivity of those pathogens to humans. The relative values of QMRA and epidemiology are strongly debated (Haas and Eisenberg, 2001).

### *3.6.1 The mathematical modelling of health risk*

Establishing the exposure setting is the first step to the mathematical evaluation of microbial risk. The purpose is to determine the possible pathogens present, dose(s) consumed and the characteristics of the pathogen(s) that will define the outcome.

The quantitative approach to microbial risk assessment is based on the chemical risk assessment paradigm, and has been reviewed by Haas *et al.* (1999). As with chemical risk assessment, this is a formalised procedure involving four key steps (Table 3.3), each of which is briefly described below.

**Table 3.3.  Steps involved in quantitative microbial risk assessment**

(Adapted from National Research Council, 1983)

| Step | Aim |
|---|---|
| 1.  Problem formalisation and hazard identification | To describe the overall environmental setting and relevant pathogens that may cause acute or chronic effects to human health. |
| 2.  Dose-response analysis | To find appropriate relationship(s) between pathogen exposure and infection or illness (from epidemiological studies). |
| 3.  Exposure assessment | To determine the size and nature of the populations exposed to each identified pathogen by route, amount and duration of the exposure. |
| 4.  Risk characterisation | To integrate the information from exposure and dose-response, to express public health outcomes, taking into account the variability and uncertainty of the estimations. |

While the conceptual framework for both chemical and microbial risk assessments is the same, pathogens differ from toxic chemicals in several key ways:

- The variability of different strains of the one pathogen to cause disease (differing virulence).

- Virulence can evolve as the pathogen passes through various infected individuals.

- Pathogens are generally not evenly suspended in water.

- Pathogens can be passed from one person to many (secondary spread), from either healthy but infected (asymptomatic) or ill (symptomatic) hosts.

- Whether a person becomes infected or ill depends not only on the health of the person, but also on their pre-existing immunity and pathogen dose.

### 3.6.2    Hazard identification (hazard assessment)

Pathogenic microorganisms are relatively well described in the scientific literature and, apart from emerging waterborne pathogens (LeChevallier *et al.*, 1999a,b), data on their characteristics are generally available. Data needed for the risk assessment process include the severity of the outcome, susceptibility (long and short-term immunity), susceptible populations and secondary (person-to-person) disease transmission. The outcomes of the exposure include non-infection, asymptomatic infection and various levels of morbidity and mortality. Gender, age and some forms of susceptibility may also affect the outcome. Severe morbidity or mortality resulting from waterborne exposures are significant in developing countries, but are relatively rare in industrialised countries.

### 3.6.2.1   Outbreaks

To properly conduct risk assessment, the hazard must be identified and outbreaks provide important data in microbial risk assessment. The pathogen responsible for the outbreak must be identified, the severity and contagiousness of the infection can be described, the patterns of transmission in the population can be studied and control measures can be evaluated. Waterborne disease outbreak surveillance is key to this evaluation, and identification of the aetiologic agent is dependent on the timely recognition of the outbreak, so that appropriate clinical and environment samples can be obtained. The interests and expertise of investigators and the routine practices of local laboratories can also influence whether the aetiologic agent is identified (Frost *et al.*, 1996). Diarrhoeal stool specimens, for example, are generally examined for bacterial pathogens, but not for viruses. In most laboratories, testing for *Cryptosporidium* is only undertaken if requested and is not included in routine stool examinations for ova and other parasites. Hence, it is not surprising that even in the USA, with one of the most comprehensive registers of waterborne outbreaks, between 1992-1996 the causative organism was not identified in over 40% of investigations (Levy *et al.*, 1998).

The water quality data collected during and/or before the outbreak can be useful in identifying the causes of the outbreak and in preventing their reoccurrence. (Methods used for microbial water quality assessment are discussed in Chapter 8 and their use in outbreak investigation is described in Chapter 7). While background data on the level of faecal contamination, if not sewage pollution in water is very valuable, care is needed in interpreting data on finding or not finding pathogens. In particular, molecular epidemiology or similar typing methods are necessary to confirm if the species identified from water was also the agent present in the infected host (Chapter 7). There has been considerable controversy over a number of species of opportunistic bacterial pathogens with apparently non-pathogenic strains that may be found in drinking water, versus different strains (and presumably non-water sources) causing illness (Edberg *et al.*, 1986; Havelaar *et al.*, 1992; Kühn *et al.*, 1997).

### 3.6.2.2 'Emerging' pathogens

As new pathogens are being described in the population or in the environment, their potential for being transmitted by the water route must be evaluated. Basic characteristics that allow a pathogen to be waterborne include:

- Excretion in the faeces and/or urine.

- An environmentally persistent stage.

- The ability to cause infection when inhaled or ingested.

Emerging pathogens include those that are increasingly being recognised as important contributors to waterborne disease as well as those that are newly discovered. As such, they include:

- Viruses: new enteroviruses, human caliciviruses (including Norwalk-like viruses), and hepatitis E.

- Parasitic protozoa: *Cyclospora cayetanensis*, various microsporidia and *Toxoplasma gondii*.

- Bacteria: Mycobacterium avium complex, Helicobacter pylori, pathogenic Escherichia coli and Campylobacter jejuni (LeChevallier et al., 1999a,b).

- Toxic cyanobacteria (Chorus and Bartram, 1999).

- Most faecal-oral pathogens are identified as causing acute gastrointestinal illnesses, with the major exceptions being hepatitis A and E viruses, *Helicobacter pylori*, *Salmonella typhi* and hookworm infection. However, it is important to note (as mentioned in Chapter 1) that some commonly recognised diseases (such as arthritis, type 1 diabetes mellitus, abortion, Guillain-Barré and Miller Fisher Syndrome) have been associated with, or are suspected to be caused by, infection with viral or bacterial pathogens excreted by humans or animals (Duim *et al.*, 2000; Frisk *et al.*, 1992; Gurgan and Diker, 1994; Havelaar *et al.*, 2000; Maki-Ikola and Granfors, 1992; Niklasson *et al.*, 1998).

### 3.6.3 *Dose-response analysis*

For QMRA, human dose-response studies are available for a few pathogens and can be used to estimate the effects of low level exposure to these microorganisms (Haas and Eisenberg, 2001). Two models of the infection process have been proposed: the exponential model (Equation 1) and the beta-Poisson model (Equation 2). These have been developed from biologically plausible assumptions about the infection process. Models may fit available data in a statistically acceptable sense and yet provide very different estimates for the risk at an extrapolated low dose; a situation that has frequently caused argument in chemical risk assessment. In QMRA, it may be possible to test the potential appropriateness of different dose-response functions by validating with outbreak data (Eisenberg *et al.*, 1998).

Exponential model:

$$Probability_{infection} = 1 - \exp(-rD) \qquad \textit{Equation 1}$$

Where D = pathogen dose; r = fraction of pathogens that survives to produce an infection.

Beta-Poisson model:

$$Probability_{infection} = 1 - (1 + (D/ID_{50}))^{-\alpha} \qquad \textit{Equation 2}$$

Where D = pathogen dose; $\alpha$ and $ID_{50}$ are parameters of the beta-distribution used to describe variability in survival.

Given a set of dose-response data, *i.e.* exposure of populations to various doses of microorganisms and measurement of response (such as infection), the best fitting parameters of a dose-response relationship may be computed via standard maximum likelihood techniques. The method has been used for human viruses, parasitic protozoa and some bacterial pathogens (Haas *et al.*, 1999). Confidence limits to the parameters can then be estimated, and used as a basis for low-dose extrapolation (Kang *et al.*, 2000). It should be noted that, in general, dose-response studies have been conducted on healthy adults and may not reflect the response of the general population or of more susceptible population segments.

During an outbreak, individuals are exposed to different levels of the pathogen(s): the volume of water ingested may be coupled with data on the level of contamination of the water. These data can provide a dose-response relationship confirming volunteer studies. Furthermore, information on susceptible sub-populations (such as children and the immuno-compromised) may also be forthcoming. For example, waterborne outbreaks of cryptosporidiosis indicate that the general population may contract watery diarrhoea that lasts up to several days, whereas HIV-patients may be untreatable and die, thereby creating a much more significant health burden if the latter are included in a risk assessment (Perz *et al.*, 1998).

Volunteer feeding studies have provided data on the dose-response curve for several pathogens (Haas *et al.*, 1999). It is, however, often difficult to obtain data on low doses as large numbers of volunteers would be needed to define the lower bounds of the dose-response curve. It is also difficult to extrapolate from a single strain to give a generalised model for a pathogen. Virulence differs from one strain to another and the outcomes are often very different (*e.g. E. coli* enteropathogenic versus non-enteropathogenic strains). The debate around the human health significance of exposure to human versus animal strains of *Cryptosporidium parvum* is another example. Feeding trials with three different

91

bovine strains of *C. parvum* have generated 50% infective doses ($ID_{50}$) for oocysts in healthy human volunteers ranging between 9 and 1 042 (Okhuysen *et al.*, 1999). Such a wide range is potentially problematic as the $ID_{50}$ is the parameter defining the slope of the dose-response curve in the beta-Poisson model. A further complication is that pre-existing immunity may provide protection from infection and illness at low oocyst doses (Chappell *et al.*, 1999), thereby changing the low dose-response extrapolation in a manner not accounted for by any current model.

Relatively few data points are used to generate the curve and the degree of uncertainty over the position of each data point is high. Each data point is a sample mean of the probability of illness for people exposed to a set dose of pathogen. The confidence intervals for each sample mean will be wide. It is unlikely that all the measured points exactly correspond with the true population means for each dose. In such circumstances it is impossible to be certain about what dose-response model would best fit the actual curve (as opposed to the curve of the sample means). There is, therefore, considerable uncertainty in which model best fits the actual dose response curve and what its parameters should be (Coleman and Marks, 1998). The impact of these uncertainties is most marked at low doses (*i.e.* at the dose that will most frequently be experienced in real life). Therefore, the predicted number of illnesses following low dose exposure can vary by several orders of magnitude (Holcomb *et al.*, 1999).

### 3.6.4    Exposure assessment

The actual dose consumed by an individual is generally unknown and difficult to estimate. Methods for the detection of some pathogens are not even available, and most pathogens occur at very low levels in treated water (generally below detection). The general level of some pathogens (*e.g.* enteroviruses, *Giardia*, *Cryptosporidium*), however, are available for sewage and untreated water. These raw water values can be used, along with the proportion of surrogate removed by treatment, to indirectly estimate the level of individual pathogens after treatment, thereby providing an estimate of the 'dose' in the water. The possible uses of surrogates and indicators are further discussed below.

For drinking waters, the volume ingested per 'exposure' is relatively well defined after several studies in a number of countries (*e.g.* Roseberry and Burmaster, 1992). A volume of two litres per person per day is often used to estimate drinking water exposure, but this does not reflect the fact that only a fraction of that volume is consumed unmodified (especially unboiled). This is important for QMRA as microorganisms are inactivated by heat; therefore water consumed in hot drinks or used in the preparation of cooked food would not be a risk factor.

Viruses and parasites have been detected in drinking water, which was otherwise apparently safe, without any detectable health effect being seen in the receiving population (Bouchier, 1998). Possible reasons for this include false positive detections, the presence of non-infective pathogens and the pathogen is present in a concentration below that which would be expected to cause detectable disease in the population. On the other hand, unrealistically large volumes of drinking water would need to be sampled for example to meet the USEPA's level of acceptable waterborne risk ($<10^{-4}$ infections per annum – see 1.5.1). Translating this for *Cryptosporidium parvum* would mean that 500 samples of 2 000 litres each would be needed to make a reasonably accurate estimation of the allowed concentration ($7 \times 10^{-6}$ per litre) (Teunis *et al.*, 1996). Furthermore, depending on the detection method used, an unknown proportion of pathogens isolated from the environment may be incapable of causing infection. Therefore, alternative strategies are recommended to estimate pathogen concentrations.

The applications of coliform bacteria to index the pollution of source water, or as an indicator of water treatment efficacy or recontamination of treated water have provided little information on health effects in developed regions. Nonetheless, these organisms can play an important part in estimating pathogen numbers for a screening-level or first tier of a QMRA. For example, direct measurement of viral, parasitic protozoa and bacterial pathogens is possible for sewage effluents, as is the estimation of pathogen prevalence data for the faeces of some domestic animals. Hence, predictions of pathogens in source waters can be made if the relative proportion of human and animal faecal load is determined by, say, the analysis of faecal sterols (Leeming *et al.*, 1998). For environments where sewage is the primary faecal contaminant, then pathogen dilutions in source waters can be estimated directly by the dilution of thermotolerant coliforms (index for bacterial pathogen contamination) and spores of *Clostridium perfringens* (index for the hardier viral and protozoan pathogens) (Medema *et al.*, 1997).

For physical treatment barriers, such as sand or membrane filters, and for disinfection by chlorine, ozone or UV, surrogates for pathogen removal are also generally accepted. Total aerobic spores or spores of *C. perfringens* are reasonable surrogates for the cysts and oocysts of parasitic protozoa and coliphages may also be appropriate for human enteric viruses (Facile *et al.*, 2000; Hijnen *et al.*, 2000; Ndiongue *et al.*, 2000; Owens *et al.*, 2000). Note that while coliphages make good models for human virus removal by physical means, that may not be the case for mixed oxidants (Casteel *et al.*, 2000).

### 3.6.5    Infectious disease models and risk characterisation

As outlined in the previous sections, attempts to provide a quantitative assessment of human health risks associated with the ingestion of waterborne pathogens have generally focused on static models that calculate the probability of individual infection or disease as a result of a single exposure event. They do not address the properties that are unique to infectious disease transmission such as secondary transmission, immunity and population dynamics (Haas and Eisenberg, 2001). To understand the role that water plays in the transmission of enteric pathogens and to estimate the risk of disease due to drinking water within a defined population it is important to study the complete disease transmission system, as illustrated in Figure 3.2. It is also important to recognise the additional pathways that describe the natural history of enteric pathogens: animal-to-environment-person, person-to-environment-to-person, and person-to-person (Eisenberg *et al.,* 2001).

A fundamental concept in disease transmission models is the reproduction number, $R_0$, which is defined as the number of infections that result from the introduction of one index case into a population of susceptible individuals. Therefore, $R_0$ is a measure of the ability of a pathogen to move through a population. An $R_0$ >1 suggests that the pathogen is multiplying within a community and that prevalence is increasing, whereas an $R_0$ <1 suggests that the disease is dying out of the population. An $R_0$ that is on average equal to 1 suggests that the disease is endemic in the population. There are various methods to estimate $R_0$ for different pathogens and in different environmental settings (Dietz, 1993). Measles, for example, is a highly infectious respiratory transmitted disease and has been estimated to have an $R_0$ of approximately 14. Polio, on the other hand, a waterborne pathogen has an $R_0$ of approximately 6.

**Figure 3.2.  Conceptual model for rotavirus infection pathways**

(from Haas and Eisenberg, 2001)

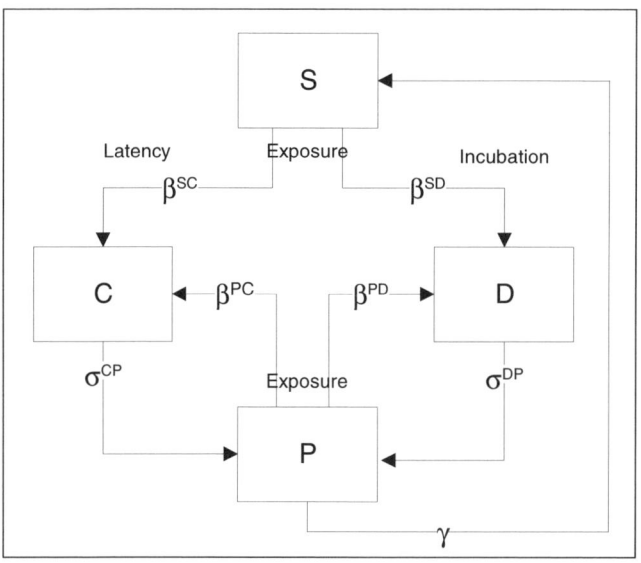

S: susceptible = not infectious, not symptomatic. C: carrier = infectious, not symptomatic. D: diseased = infectious, symptomatic. P: post Infection = not infectious, not symptomatic with short-term or partial immunity.

Summarising the previous sections, the individual daily dose of pathogenic microorganisms via some particular product may be calculated as (Teunis *et al.*, 1996):

$$Dose = C \times \frac{1}{R} \times I \times 10^{-DR} \times V \qquad\qquad Equation\ 3$$

$C$ = Concentration of pathogenic microorganisms in raw (source) materials (or partially processed products, if data are available).

$R$ = Recovery of the detection method.

$I$ = Fraction of the detected pathogens that is capable of infection (viability).

$DR$ = Removal or inactivation efficiency of the treatment process, expressed as its Decimal Reduction factor ($DR = 0$ when concentrations in the finished product are available).

$V$ = Daily individual consumption of the considered product.

In many cases, risk evaluations start from the assumption that the dose-response relationship is approximately linear at low doses. Therefore, at very low doses, calculation of the risk of infection simply consists of multiplying the dose estimate with the slope of the dose-response relationship. Estimates of daily risk may be extrapolated to yearly risk. When $P_1^*$ and $P_n^*$ are the probabilities of infection after a single (*e.g.* daily) exposure and after repeated exposures (n times a daily exposure) respectively:

$$P_n^* = 1 - (1 - P_1^*)^n \approx n \times P_1^* \qquad\qquad Equation\ 4$$

The latter simplification is valid as long as $P_1^* \ll 1$ (Haas *et al.*, 1999).

From the above discussions it would seem that microbial data, whether relating to indicator parameters or pathogens, have most relevance to the exposure assessment phase of QMRA. These provide estimates of actual levels of pathogens in water or the likelihood that water is exposed to faecal pollution. However, caution must be exercised in assuming a direct relationship between this level and risk to health. Despite the use of numbers and mathematical equations, QMRA is not yet an exact science.

## 3.7    Qualitative risk assessment

Qualitative methods for analysing microbial hazards and managing risks are commonplace within the food industry. They are applied as part of a systematic process including Hazard Analysis Critical Control Points (HACCP) (Coleman and Marks, 1999), which has recently been taken up by the water industry (Havelaar, 1994; Barry *et al.*, 1998; Deere and Davison,1998; Gray and Morain, 2000; Deere *et al.*, 2001; Dewettinck *et al.*, 2001; Davison *et al.*, 2002).

Although hazard identification and exposure assessment are common issues across qualitative and quantitative methods, dose-response models and risk characterisation steps (Table 3.3), are usually replaced with risk rankings in qualitative assessments. These rankings are generally derived from expert opinion summarising:

- Likelihood of possible risk pathways.

- Severity of outcome from each pathway.

- Numbers of people that may be impacted.

Water agencies are now focusing on the whole system approach, as illustrated in Figure 3.3, which includes an assessment of all types of physical, chemical and microbiological risks. Possible ranking schemes are numerous, but follow the generic structure indicated in Table 3.4, with Table 3.5 illustrating a simple risk scoring table.

**Table 3.4. Possible qualitative risk assessment approach to rank or scale hazardous scenarios**

| Step | Comment |
|------|---------|
| 1. Hazard scenario | Identification of hazardous scenarios, such as massive rainfall-induced contamination of source water, filter breakthrough or loss/breakdown of chemical disinfection system (*i.e.* not necessarily limited to a single pathogen). |
| 2. Likelihood | Ranking or scaling of how likely the event is (*e.g.* # events per year). |
| 3. Consequence | Ranking or scaling of the consequence (*e.g.* short-term injury or ill-health through to permanent disability or death). |
| 4. Scale of effect | Consideration of the number of people affected by the hazard scenario. |
| 5. Risk score | Different weightings may be given to (2) to (3) and multiplied to give a value for each hazard scenario. |
| 6. Rank | Each hazard scenario is then ranked, to provide a priority list for risk management. |

**Table 3.5. Simple risk scoring table for prioritising risks**

(Davison *et al.*, 2002)

| Likelihood | Severity of consequences | | | | |
|------------|--------------------------|-------|----------|-------|--------------|
|            | Insignificant | Minor | Moderate | Major | Catastrophic |
| Almost certain | 5 | 10 | 15 | 20 | 25 |
| Likely | 4 | 8 | 12 | 16 | 20 |
| Moderate | 3 | 6 | 9 | 12 | 15 |
| Unlikely | 2 | 4 | 6 | 8 | 10 |
| Rare | 1 | 2 | 3 | 4 | 5 |

The risk score for a particular hazard = likelihood × severity of consequences.

An example of the descriptive terms that can be used to rate the likelihood and severity for calculation of the risk score is given in Table 3.6.

**Figure 3.3. Generic flow diagram for sources of microbial risk in a drinking water context**

(Adapted from Stevens *et al.*, 1995)

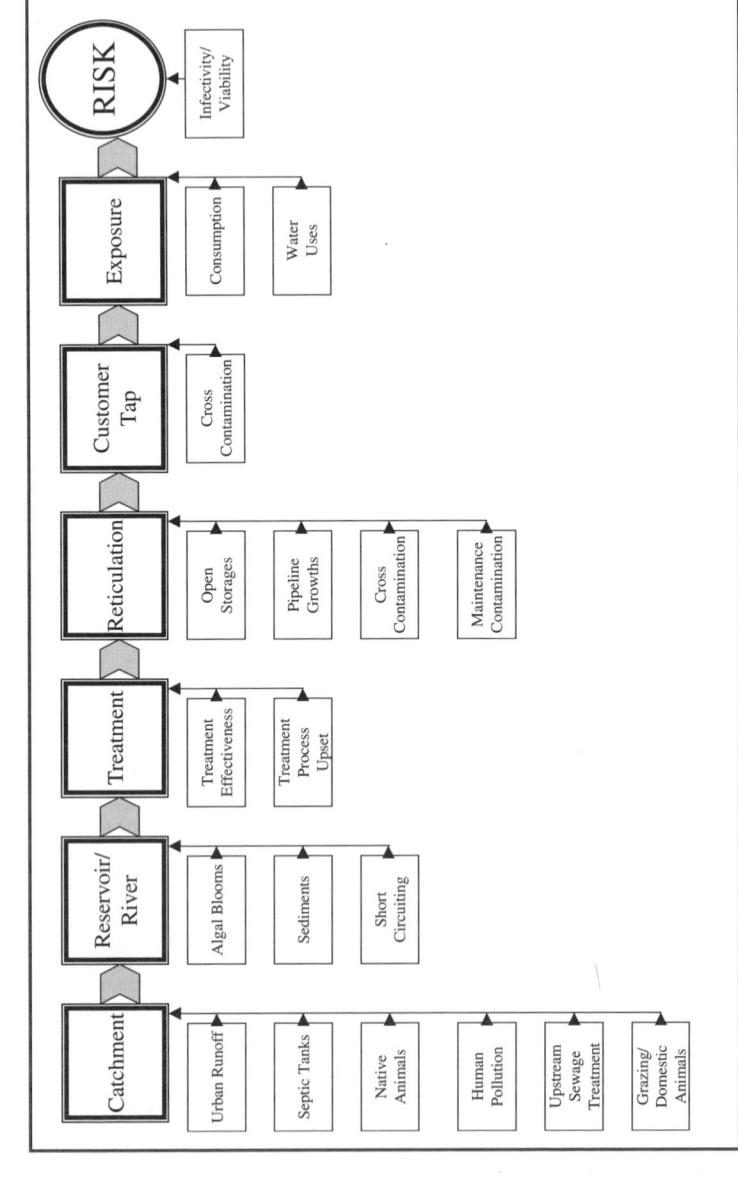

**Table 3.6. Example descriptive terms for risk score calculation**

(Davison *et al.*, 2002)

| Item | Definition | Weighting |
|------|------------|-----------|
| Almost certain | Once a day | 5 |
| Likely | Once per week | 4 |
| Moderate | Once per month | 3 |
| Unlikely | Once per year | 2 |
| Rare | Once every five years | 1 |
| Catastrophic | Potentially lethal to large population | 5 |
| Major | Potentially lethal to small population | 4 |
| Moderate | Potentially harmful to large population | 3 |
| Minor | Potentially harmful to small population | 2 |
| Insignificant | No impact or not detectable | 1 |

Compared to both epidemiological and quantitative microbial risk assessment, this approach does not seek to determine actual levels of disease associated with a supply. As such, criticisms cannot be made that the conclusions are imprecise compared with reality. The other advantage that this approach has over other methods is that out of the process itself solutions to minimise risk will present themselves. On the other hand, reliance on 'expert opinion' does not always produce the correct answer as experts' opinions and models of the world are often subject to bias and inaccuracies as with any other source of data (Hunter and Fewtrell, 2001).

### 3.7.1 *Indicators and qualitative microbial risk assessment*

Microbial and other indicator analyses will be a major source of evidence at several stages of qualitative risk assessment. The role of such information in conducting assessments of source water quality, treatment efficacy and integrity of the distribution system are discussed in more details in Chapters 4–6.

As will be seen, studies on the presence of indicator organisms frequently provide more useful information for qualitative risk assessment than do studies on enumeration of specific pathogens. Nevertheless, well-designed studies of specific pathogens can also be of great value in certain situation. For example, the detection of *E. coli*, faecal streptococci or sulphite-reducing *Clostridia* in source water all indicate that the water is subject to contamination from human or animal faeces. Detection and typing of *Cryptosporidium* in source water will give a better understanding of the risk to the water supply system and the sources of contamination.

Coliform bacteria in treated water may give an indication that water treatment systems are not operating satisfactorily or that water is becoming contaminated within the distribution system. However, coliform bacteria alone are not good indicators of risk from chlorine-resistant pathogens such as *Cryptosporidium*. Some indicator organisms may be naturally present in the source water or can be deliberately seeded into the inlet to a water treatment works and monitored at various stages in the treatment and distribution in order to demonstrate the effectiveness of the whole system.

## 3.8    Summary

Microbial and other indicator parameters play an essential role in all the models used in the assessment of risk discussed in this chapter. However, the exact relationship between these indicator parameters and risk to health is still far from clear. Although studies have shown that turbidity and faecal streptococci are independent indicators of health risk there is no clear-cut predictive relationship. Even where information on pathogens in potable water is available, current quantitative risk assessment models have considerable uncertainty in their calculated risk. Perhaps the real value of such indicator parameters is in qualitative risk assessment where they can be used for identifying where failures may occur in the water extraction, treatment and distribution system.

# REFERENCES

Andersson, Y. and Bohan, P. (2001) Disease surveillance and waterborne outbreaks. In: *Water Quality: Guidelines, Standards and Health. Assessment of Risk and Risk management for Water-related Infectious Disease.* Fewtrell, L. and Bartram, J. (Eds.) IWA Publishing, London. pp. 115-133.

Barry, S.J., Atwill, E.R., Tate, K.W., Koopman, T.S., Cullor, J. and Huff, T. (1998) Developing and implementing a HACCP-based programme to control *Cryptosporidium* and other waterborne pathogens in the Alameda Creek watershed: Case study. American Water Works Association Annual Conference, 21-25 June 1998, Dallas. *Texas Water Resources* Vol. **B**, 57-69.

Batik, O., Craun, G.F. and Pipes, W.O. (1983) Routine coliform monitoring and water-borne disease outbreaks. *Journal of Environmental Health* **45**, 227-230.

Batik, O., Craun, G.F., Tuthil, R.W. and Kroemer, D.F. (1979) An epidemiologic study of the relationship between hepatitis A and water supply characteristics and treatment. *American Journal of Public Health* **70**, 167-169.

Beaudeau, P., Payment, P., Bourderont, D., Mansotte, F., Boudhabay, O., Laubiès, B. and Verdière, J. (1999) A time series study of anti-diarrheal drug sales and tap-water quality. *International Journal of Environmental Health Research* **9**(4), 293-312.

Blum, D. and Feachem, R.G. (1983) Measuring the impact of water supply and sanitation investments on diarrhoeal diseases: problems of methodology. *International Journal of Epidemiology* **12**(3), 357-365.

Blumenthal, U.J., Fleisher, J.M., Esrey, S.A. and Peasey, A. (2001) Epidemiology: a tool for the assessment of risk. In: *Water Quality: Guidelines, Standards and Health. Assessment of risk and risk*

101

*management for water-related infectious disease*. Fewtrell, L. and Bartram, J. (Eds.). IWA Publishing, London. pp135-160.

Bouchier, I. (1998) *Cryptosporidium in Water Supplies*. Third Report of the Group of Experts to Department of the Environment, Transport and the Regions and Department of Health. November, 1998. Drinking Water Inspectorate, London. http://www.dwi.detr.gov.uk/pubs/bouchier/index.htm pages.

Casteel, M.J., Sobsey, M.D. and Arrowood, M.J. (2000) Inactivation of *Cryptosporidium parvum* oocysts and other microbes in water and wastewater by electrochemically generated mixed oxidants. *Water Science and Technology* **41**(7), 127-134.

Chappell, C.L., Okhuysen, P.C., Sterling, C.R., Wang, C., Jakubowski, W. and DuPont, H.L. (1999) Infectivity of *Cryptosporidium parvum* in healthy adults with pre-existing anti-*C. parvum* serum immunoglobulin G. *American Journal of Tropical Medicine and Hygiene* **60**(1),157-164.

Chorus, I. and Bartram, J. (Eds.) (1999) *Toxic Cyanobacteria in Water*. E&FN Spon, London.

Coleman, M. and Marks, H. (1998) Topics in dose-response modelling. *Journal of Food Protection* **61**, 1550-1559.

Coleman, M.E. and Marks, H.M. (1999) Qualitative and quantitative risk assessment. *Food Control* **10**, 289-297.

Collin, J.F., Milet, J.J., Morlot, M. and Foliguet, J.M. (1981) Eau d'adduction et gastroentérites en Meurthe-et-Moselle. *J. Franc. Hydrologie* **12**, 155-174.

Craun, G.F., Berger, P.S. and Calderon, R.L. (1997) Coliform bacteria and waterborne disease outbreaks. *Journal of the American Water Works Association* **89**(3), 96-104.

Davison, A., Howard, G., Stevens, M., Callan, P., Kirby, R., Deere, D. and Bartram, J. (2002) *Water Safety Plans*. WHO/SHE/WSH/02/09 World Health Organization, Geneva, Switzerland.

Deere, D., Stevens, M., Davison, A., Helm, G. and Dufour, A. (2001) Management Strategies. In: *Water Quality: Guidelines, Standards and Health. Assessment of risk and risk management for water-related infectious disease*. Fewtrell, L. and Bartram, J. (Eds.) IWA Publishing, London. pp. 257-288.

Deere, D.A. and Davison, A.D. (1998) Safe drinking water. Are food guidelines the answer? *Water* **25**, 21-24.

Dewettinck, T., Van Houtte, E., Geenens, D., Van Hege, K. and Verstraete, W. (2001) HACCP (Hazard Analysis and Critical Control Points) to guarantee safe water reuse and drinking water production – a case study. *Water Science and Technology* **43**(12), 31-38.

Dietz, K. (1993) The estimation of the basic reproduction number for infectious diseases. *Statistical Methods in Medical Research* **2**, 23-41.

Duim, B., Ang, C.W, van Belkum, A., Rigter, A., van Leeuwen, N.W.J., Endtz, H.P. and Wagenaar, J.A. (2000) Amplified fragment length polymorphism analysis of *Campylobacter jejuni* strains isolated from chickens and from patients with gastroenteritis or Guillain-Barré or Miller Fisher Syndrome. *Applied and Environmental Microbiology* **66**(9), 3917-3923.

Edberg, S.L., Pisticelli, V. and Cartter, M. (1986) Phenotypic characteristics of coliform and non coliform bacteria from a public water supply compared with regional and national clinical species. *Applied and Environmental Microbiology* **52**, 474-478.

Eisenberg, J.N.S., Bartram, J. and Hunter, P.R. (2001) A public health perspective for establishing water-related guidelines and standards. In: *Water Quality: Guidelines, Standards and Health. Assessment of risk and risk management for water-related infectious disease*. Fewtrell, L. and Bartram, J. (Eds.) IWA Publishing, London. pp. 229-256.

Eisenberg, J.N.S., Seto, E.Y.W., Colford, J., Olivieri, A.W. and Spear, R.C. (1998) An analysis of the Milwaukee *Cryptosporidium* outbreak based on a dynamic model of disease transmission. *Epidemiology* **9**(3), 255-263.

Esrey, S.A., Feachem, R.G. and Hughes, J.M. (1985) Interventions for the control of diarrhoeal diseases among young children: improving water supplies and excreta disposal facilities. *Bulletin of the World Health Organization* **63**(4), 757-772.

Esrey, S.A., Potash, J.B., Roberts, L. and Schiff, C. (1991) Effects of improved water supply and sanitation on ascariasis, diarrhoea, dracunculiasis, hookworm infection, schistosomiasis and trachoma. *Bulletin of the World Health Organization* **69**(5), 609-621.

Facile, N., Barbeau, B., Prévost, M. and Koudjonou, B. (2000) Evaluating bacterial aerobic spores as a surrogate for *Giardia* and *Cryptosporidium* inactivation by ozone. *Water Research* **34**(12), 3238-3246.

Fattal, B., Guttman-Bass, N., Agursky, T. and Shuval, H.I. (1988) Evaluation of health risk associated with drinking water quality in agricultural communities. *Water Science and Technology* **20**, 409-415.

Ferley, J.P., Zmirou, D., Collin, J.F. and Charrel, M. (1986) Etude longitudinale des risques liés à la consommation d' eaux non conformes aux normes bactériologiques. *Rev. Epidemiol. Sante Publique* **34**, 89-99.

Frisk, G., Nilsson, E., Tuvemo, T., Friman, G. and Diderholm, H. (1992) The possible role of coxsackie A and echo viruses in the pathogenesis of type 1 diabetes mellitus studied by IgM analysis. *Journal of Infection* **24**, 13-22.

Frost, F.J., Craun, G.F. and Calderon, R.L. (1996) Waterborne disease surveillance. *Journal of the American Water Works Association* **88**(9), 66-75.

Gordis, L. (2000) *Epidemiology 2$^{nd}$ Edition*. W.B Saunders Company, Philadelphia.

Gostin, L.O., Lazzarini, Z., Neslund, V.S. and Osterholm, M.T. (2000) Water quality laws and waterborne diseases: *Cryptosporidium* and other emerging pathogens. *American Journal of Public Health* **90**(6), 847-853.

Gray, R. and Morain, M. (2000) HACCP application to Brisbane Water. *Water* **27**, 41-42.

Gurgan, T. and Diker, K.S. (1994) Abortion associated with *Campylobacter upsaliensis. Journal of Clinical Microbiology* **32**, 3093-3094.

Haas, C. and Eisenberg, J, N.S. (2001) Risk assessment. In: *Water Quality: Guidelines, Standards and Health. Assessment of risk and risk management for water-related infectious disease.* Fewtrell, L. and Bartram, J. (Eds.) IWA Publishing, London. pp. 161-183.

Haas, C.N., Rose, J.B. and Gerba, C.P. (1999) *Quantitative Microbial Risk Assessment*. John Wiley, New York.

Havelaar, A.H. (1994) Application of HACCP to drinking water supply. *Food Control* **5**, 145-152.

Havelaar, A.H., de Wit, M.A.S. and van Koningsveld, R. (2000) Healthburden in the Netherlands (1990-1995) due to infections with thermophilic *Campylobacter* species. Report no. 284550 004. RIVM (National Institute of Public Health and the Environment), Bilthoven.

Havelaar, A.H., Schets, F.M., van Silfhout, A., Jansen, W.H., Wieten, G. and van der Kooij, D. (1992) Typing of *Aeromonas* strains from patients with diarrhoea and from drinking water. *Journal of Applied Bacteriology* **72**(5), 435-444.

Hijnen, W.A.M., Willemsen-Zwaagstra, J., Hiemstra, P., Medema, G.J. and van der Kooij, D. (2000) Removal of sulphite-reducing clostridia spores by full-scale water treatment processes as a surrogate for protozoan (oo)cysts removal. *Water Science and Technology* **41**(7), 165-171.

Holcomb, D.L., Smith, M.A., Ware, M.A., Hung, Y.-C., Brackett, R.E. and Doyle, M.P. (1999) Comparison of six dose-response models for use with food-borne pathogens. *Risk Analysis* **19**, 1091-1100.

Hunter, P.R. (2000) Modelling the impact of prior immunity, case misclassification and bias on case-control studies in the investigation of outbreaks of cryptosporidiosis. *Epidemiology and Infection* **125**, 713-718.

Hunter, P.R. and Fewtrell, L. (2001) Acceptable risk. In: *Water Quality: Guidelines, Standards and Health. Assessment of risk and risk management for water-related infectious disease.* Fewtrell, L. and Bartram, J. (Eds.) IWA Publishing, London. pp.207-227.

Hunter, P.R. and Syed, Q. (2001) Community surveys of self-reported diarrhoea can dramatically overestimate the size of outbreaks of waterborne cryptosporidiosis. *Water Science and Technology* **43**(12), 27-30.

Isaäcson, M. and Sayed, A.R. (1988) Health aspects of the use of recycled water in Windhoek, SWA/Namibia, 1974-1983. Diarrhoeal diseases and the consumption of reclaimed water. *South Africa Medical Journal* **7**, 596-9.

Kang, S.H., Kodell, R.L. and Chen, J.J. (2000) Incorporating model uncertainties along with data uncertainties in microbial risk assessment. *Regulatory Toxicology and Pharmacology* **32**(1), 68-72.

Kühn, I., Albert, M.J., Ansaruzzaman, M., Bhuiyan, N.A., Alabi, S.A., Islam, M.S., Neogi, P.K.B., Huys, G., Janssen, P., Kersters, K. and Möllby, R. (1997) Characterization of *Aeromonas* spp. isolated from humans with diarrhea, from healthy controls, and from surface water in Bangladesh. *Journal of Clinical Microbiology* **35**(2), 369-373.

LeChevallier, M.W., Abbaszadegan, M., Camper, A.K., Hurst, C.J., Izaguirre, G., Marshall, M.M., Nauovitz, D., Payment, P., Rice, E.W., Rose, J., Schaub, S., Slifko, T.R., Smith, D.B., Smith, H.V., Sterling, C.R. and Stewart, M. (1999a) Committee report: emerging pathogens - bacteria. *Journal of the American Water Works Association* **91**(9), 101-109.

LeChevallier, M.W., Abbaszadegan, M., Camper, A.K., Hurst, C.J., Izaguirre, G., Marshall, M.M., Nauovitz, D., Payment, P., Rice, E.W., Rose, J., Schaub, S., Slifko, T.R., Smith, D.B., Smith, H.V., Sterling, C.R. and Stewart, M. (1999b) Committee report: emerging pathogens - viruses, protozoa, and algal toxins. *Journal of the American Water Works Association* **91**(9), 110-121.

Leeming, R., Bate, N., Hewlett, R. and Nichols, P.D. (1998) Discriminating faecal pollution: a case study of stormwater entering Port Philip Bay, Australia. *Water Science and Technology* **38**(10), 15-22.

Levy, D.A., Bens, M.S., Craun, G.F., Calderon, R.L. and Herwaldt, B.L. (1998) Surveillance for waterborne-disease outbreaks -- United States, 1995-1996. *Morbidity and Mortality Weekly Report* **47**(SS-5), 1-34.

Macgill, S., Fewtrell, L., Chudley, J. and Kay, D. (2001) Quality audit and the assessment of waterborne risk. In: *Water Quality: Guidelines, Standards and Health. Assessment of risk and risk management for water-related infectious disease.* Fewtrell, L. and Bartram, J. (Eds.) IWA Publishing, London. pp. 185-206.

MacKenzie, W.R., Hoxie, N.J., Proctor, M.E., Gradus, M.S., Blair, K.A., Peterson, D.E., Kazmierczak, J.J., Addiss, D.G., Fox, K.R., Rose, J.B. and Davis, J.P. (1994) A massive outbreak in Milwaukee of *Cryptosporidium* infection transmitted through the public water supply. *New England Journal of Medicine* **331**, 161-167.

Maki-Ikola, O. and Granfors, K. (1992) *Salmonella*-triggered reactive arthritis. *Lancet* **339**, 1096-1098.

Medema, G.J., Bahar, M. and Schets, F.M. (1997) Survival of *Cryptosporidium parvum*, *Escherichia coli*, faecal enterococci and *Clostridium perfringens* in river water - influence of temperature and autochthonous microorganisms. *Water Science and Technology* **35**(11-12), 249-252.

Morris, R.D., Naumova, E.N., Levin, R. and Munasinghe, R.L. (1996) Temporal variation in drinking water turbidity and diagnosed gastro-enteritis in Milwaukee. *American Journal of Public Health* **86**, 237-239.

National Research Council (1983) *Risk Assessment in the Federal Government: Managing the Process*. National Academy Press, Washington, DC.

Ndiongue, S., Desjardins, R. and Prévost, M. (2000) Relationships between total particle count, aerobic spore-forming bacteria and turbidity in direct filtration. *Aqua* **49**(2), 75-87.

Niklasson, B., Hörnfeldt, B. and Lundman, B. (1998) Could myocarditis, insulin-dependent diabetes mellitus and Guillain-Barré syndrome be caused by one or more infectious agents carried by rodents? *Emerging Infectious Diseases* **4**(2), 187-193.

Okhuysen, P.C., Chappell, C.L., Crabb, J.H., Sterling, C.R. and DuPont, H.L. (1999) Virulence of three distinct *Cryptosporidium parvum* isolates for healthy adults. *Journal of Infectious Diseases* **180**(4), 1275-1281.

Owens, J.H., Miltner, R.J., Rice, E.W., Johnson, C.H., Dahling, D.R., Schaefer, F.W. and Shukairy, H.M. (2000) Pilot-scale ozone inactivation of *Cryptosporidium* and other microorganisms in natural water. *Ozone-Sci. Eng.* **22**(5), 501-517.

Payment, P., Richardson, L., Siemiatycki, J., Dewar, R., Edwardes, M. and Franco, E. (1991) A randomized trial to evaluate the risk of gastrointestinal disease due to the consumption of drinking water meeting currently accepted microbiological standards. *American Journal of Public Health* **81**, 703-708.

Payment, P., Siemiatycki, J., Richardson, L., Renaud, G., Franco, E. and Prévost, M. (1997) A prospective epidemiological study of gastrointestinal health effects due to the consumption of drinking water. *International Journal of Environmental Health Research* **7**, 5-31.

Percival, S., Walker, J. and Hunter, P.R. (2000) *Microbiological Aspects of Biofilms in Drinking Water*. CRC Press, Boca Raton.

Perz, J.F., Ennever, F.K. and Leblancq, S.M. (1998) *Cryptosporidium* in tap water - comparison of predicted risks with observed levels of disease. *American Journal of Epidemiology* **147**, 289-301.

Prüss, A., Kay, D., Fewtrell, L. and Bartram, J. (2002) Estimating the burden of disease due to water, sanitation and hygiene at global level. *Environmental Health Perspectives* in press.

Quick, R.E., Venczel, L.V., Mintz, E.D., Soleto, L., Aparicio, J., Gironaz, M., Hutwagner, L., Greene, K., Bopp, C., Maloney, K., Chavez, D., Sobsey, M. and Tauxe, R.V. (1999) Diarrhoea prevention in Bolivia through point-of-use water treatment and safe storage: a promising new strategy. *Epidemiology and Infection* **122**, 83-90.

Roseberry, A.M. and Burmaster, D.E. (1992) Lognormal distributions for water intake by children and adults. *Risk Analysis* **12**(1), 99-104.

Schwartz, J., Levin, R. and Goldstein, R. (2000) Drinking water turbidity and gastrointestinal illness in the elderly of Philadelphia. *Journal of Epidemiology and Community Health* **54**(1), 45-51.

Schwartz, J., Levin, R. and Hodge, K. (1997) Drinking water turbidity and pediatric hospital use for gastrointestinal illness in Philadelphia. *Epidemiology* **8**, 615-620.

Semenza, J.C., Roberts, L., Henderson, A., Bogan, J. and Rubin, C.H. (1998) Water distribution system and diarrhoeal disease transmission: a case study in Uzbekistan. *American Journal of Tropical Medicine and Hygiene* **59**(6), 941-946.

Stevens, M., McConnell, S., Nadebaum, P.R., Chapman, M., Ananthakumar, S. and McNeil, J. (1995) Drinking water quality and treatment requirements: A risk-based approach. *Water* **22**, 12-16.

Teunis, P.F.M., van der Heijden, O.G., van der Giessen, J.W.B. and Havelaar, A.H. (1996) The dose response relation in human volunteers for gastro-intestinal pathogens. Technical report 284 550 002, RIVM (National Institute of Public Health and the Environment), Bilthoven.

Walter, S. (1991) The ecological method in the study of environmental health. II. Methodologic issues and feasibility. *Environmental Health Perspectives* **94**, 67-73.

Wheeler, J.G., Sethi, D., Cowden, J.M., Wall, P.G., Rodrigues, L.C., Tompkins, D.S., Hudson, M.J., Roderick, P.J. on behalf of the Infectious Intestinal Disease Study Executive (1999) Study of infectious intestinal disease in England: rates in the community, presenting to general practice, and reporting to national surveillance. *British Medical Journal* **318**, 1046-1050.

WHO (1993) *Guidelines for Drinking water Quality*. Volume 1: *Recommendations. Second Edition.* World Health Organisation, Geneva.

WHO (1996) *Guidelines for Drinking water Quality*. Volume 2: *Health criteria and other supporting information. Second Edition.* World Health Organisation, Geneva.

WHO (1997) *Guidelines for Drinking water Quality. Volume 3: Surveillance and control of community supplies. Second Edition.* World Health Organisation, Geneva.

Zmirou, D., Ferley, J.P., Collin, J.F., Charrel, M. and Berlin, J. (1987) A follow-up study of gastro-intestinal diseases related to bacteriologically substandard drinking water. *American Journal of Public Health* **77**, 582-584.

Zmirou, D., Rey, S., Courtois, X., Ferley, J.P., Blatier, J.F., Chevallier, P., Boudot, J., Potelon, J.L. and Mounir, R. (1995) Residual microbiological risk after simple chlorine treatment of drinking ground water in small community systems. *European Journal of Public Health* **5**, 75-81.

*Chapter 4*

# CATCHMENT CHARACTERISATION AND
# SOURCE WATER QUALITY

*G.J. Medema, S. Shaw, M. Waite, M. Snozzi, A. Morreau and W. Grabow*

## 4.1     Introduction

### *4.1.1    Select the best available source*

The first, and a key, step in providing safe drinking water is the selection of the best available source water. The most protected source waters will be the easiest and the cheapest to transform into safe drinking water. This is a general principle, and is one that has been known since the times of Plato (Whitlock, 1954). The Romans, for example, abandoned the river Tiber as a drinking water source in the third century BC and built 14 aqueducts in order to bring clean water from the surrounding hills. Principle, however, is not always translated into practice, although such oversights can have dramatic results, with a good example being provided by the Milwaukee *Cryptosporidium* outbreak (MacKenzie *et al.*, 1994), which occurred in the spring of 1993 and was estimated to have caused illness in 400 000 people. The intake of Milwaukee's Howard Avenue drinking water treatment plant was located at a site in Lake Michigan that directly received the discharge of the Milwaukee River. The city's sewage treatment plant discharged into the Milwaukee River just upstream of the river mouth. Unsurprisingly, this made the intake of the plant vulnerable to fresh faecal contamination, especially during storm events. A study in the 1960's had already shown that this lake area contained high levels of faecal pollution (Schoenen, 2001). If this information had been used to select a more appropriate location for the intake point, or to redirect the sewage discharge, the contamination level during spring 1993 would very likely have been significantly lower and perhaps a major outbreak would have been avoided.

In general, groundwater is better protected than surface water. Groundwater from deep aquifers is protected from pathogen contamination by the covering soil layers. Rain water or other water (such as from surface water infiltration, irrigation, sewer leakage etc.) that percolates through the soil can harbour pathogens but these are effectively removed by attachment to soil particles, die-off and biological processes (*e.g.* predation). Pathogen die-off during the extended time of travel from the surface through the ground to the point of abstraction in low permeability aquifers is also an important factor in reducing microbial risk. Deep groundwater from confined or semi-confined aquifers is therefore a preferred source for drinking water production. Shallower groundwater sources or groundwater that can be influenced by surface water will be more vulnerable to faecal contamination. Fine-textured soils (clay, silt) retain pathogens better than light-textured soils (sand). Soil types with a very coarse texture (fractured rock, sand and limestone, gravel) or cracks provide a relatively poor barrier against microbial contamination. Here, contact between pathogens and soil particles is less intense, leading to a lower attachment rate and greater penetration of the pathogens into the soil (REF – groundwater text).

Groundwater is not always available of suitable quality (because of salt, arsenic or fluoride content for example) or in adequate quantity. Additionally, groundwater abstraction requires drilling and pumping equipment that is not always available or sustainable especially in developing countries. Therefore, many communities rely on surface water as a source.

### 4.1.2    Catchment protection

Catchment protection is the second step in providing safe drinking water and where, for whatever reasons, source choice is limited it presents a key opportunity to minimise pathogen contamination. A major hazard to drinking water safety is presented by 'precipitation' events (rain, snowmelt), where large quantities of faecal material may be washed from the catchment into the water source, leading to the possibility of overwhelming treatment barriers and resulting in pathogen breakthrough into the finished water.

The importance of peak precipitation events is illustrated by a recent study in the USA. Rose *et al.* (2000) examined the relationship between waterborne disease outbreaks and precipitation in Pennsylvania and Colorado. The groundwater outbreaks in Pennsylvania occurred mainly in the lower river systems (Delaware, Schuylkill and Susquehanna River) where soil types are sandstone, carbonate rock and semi-consolidated sand, soil types that are vulnerable to pathogen (especially virus) penetration. The surface water outbreaks were scattered. Both groundwater and surface water outbreaks in Colorado were mostly associated with the Upper River Platte watershed and tributaries that are influenced by large cities (Denver). By correlating the time of the outbreaks with the occurrence and intensity of precipitation in the month of the outbreak or the month(s) prior to the outbreak, they associated 20 - 40% of the waterborne outbreaks in Pennsylvania and Colorado to periods of extreme precipitation (highest 10% of precipitation), for both surface water and groundwater related outbreaks. A larger scale study, which examined outbreaks and precipitation data for a 47 year period found that 51% of waterborne outbreaks (excluding those related to recreational water, cross-connection or back siphonage) were preceded by extreme precipitation (Curriero *et al.*, 2001).

Depending upon the nature of the catchment it may be possible to protect against such events by minimising possible contamination sources by, for example, removing grazing animals and diverting sewage overflows and discharge points. Where this is not feasible, a strategy for dealing with such events should be implemented. The remainder of this chapter looks at possible sources of contamination, the transport and survival of pathogens in surface and ground water and at the use of indicator parameters for informing management strategies.

## 4.2     Sources of faecal contamination

Humans, livestock and wild animals are all sources of faecal contamination, with pathogens being excreted in the faeces and occasionally urine. In general, human faecal wastes give rise to the highest risk of waterborne disease, since the probability of human pathogens being present is highest. Human enteric viruses (such as Norwalk-like caliciviruses, hepatitis A and E viruses, rotaviruses and enteroviruses) in water originate predominantly from human faecal material. Also *Shigella* spp., responsible for many waterborne disease cases and a large proportion of the deaths from waterborne disease (Traverso, 1996), is (almost) exclusively from human faecal origin. Other pathogens, such as *Campylobacter* sp., *Salmonella* spp. and *Cryptosporidium* sp., are present in both human and animal wastes. The probability of pathogens being present in these wastes depends on the presence of infected individuals that shed the pathogen.

### 4.2.1    Sources of surface water pollution

Surveys of pathogen occurrence in the sewage systems of urbanised areas show that pathogen presence in sewage and sewage effluents is the rule rather than the exception (Table 4.1). Treatment of sewage by sedimentation and activated sludge, for example, reduces the concentration of pathogens by 1-2 logs (90-99% reduction), but effluent still contains high levels of pathogens and indicator organisms. Even in (chlorine) disinfected sewage with low or no thermotolerant coliforms detectable in the effluent, viruses and protozoa are still likely to be present.

**Table 4.1. Typical concentrations of enteric pathogens and index organisms in raw and treated domestic wastewater**

| Microorganism | Raw sewage (numbers/litre) | Secondary effluent (numbers/litre) |
|---|---|---|
| **Pathogens** | | |
| Parasites | | |
| *Cryptosporidium* sp. | 1 000 – 10 000 | 10 – 1 000 |
| *Giardia* sp. | 5 000 – 50 000 | 50 – 500 |
| Viruses | | |
| Enteroviruses | 10 – 100 | 1 – 10 |
| Norwalk like viruses | 10 – 1 000 | 1 – 100 |
| Rotavirus | 10 – 100 | 1 – 10 |
| Bacteria | | |
| *Salmonella* spp. | 100 – 10 000 | 10 – 10 000 |
| **Index parameters** | | |
| Coliforms | $10^7$ - $10^9$ | $10^6$ - $10^8$ |
| Thermotolerant coliforms / *E.coli* | $10^6$ - $10^8$ | $10^5$ - $10^7$ |
| Enterococci | $10^6$ - $10^7$ | $10^4$ - $10^6$ |
| *Clostridium perfringens* | $10^5$ - $10^6$ | $10^4$ - $10^5$ |
| F-RNA phages | $10^6$ - $10^7$ | $10^5$ - $10^6$ |
| Somatic coliphages | $10^6$ - $10^7$ | $10^5$ - $10^6$ |
| Bacteroides phages | $10^4$ - $10^5$ | $10^3$ - $10^4$ |

*Source:* Rolland *et al.*, 1983; Payment *et al.*, 1986: Tartera *et al.*, 1988, 1989; Funderburg and Sorber, 1985; WRc, 1991; Havelaar *et al.*, 1986, 1993; Koenraad *et al.*, 1994; Schijven and Rijs, 2000.

Stormwater discharges are a major cause of rapid deterioration in surface water quality. Storm events bring an elevation of turbidity, suspended solids, organic matter and faecal contamination into the drainage basin, caused by urban and agricultural run-off, discharges from stormwater sewers and re-suspension of sediments. The microbiological quality of stormwater varies widely and reflects human activities in the watershed. Geldreich (1990) found that stormwater in combined sewers had more than 10-fold higher thermotolerant coliform levels ($8.9 \times 10^6$ - $4.4 \times 10^7$/l) than separate stormwater sewers ($1.0 \times 10^5$ - $3.5 \times 10^6$/l).

Livestock are a well-known source of waterborne pathogens. Several outbreaks of cryptosporidiosis in the USA, Canada and UK have been associated with the contamination of water by run-off from livestock (Craun *et al.*, 1998). At least one representative of pathogenic genera including *Cryptosporidium, Giardia, Campylobacter, Salmonella, Yersinia* and *E. coli* O157 are considered to be zoonotic. They are shed by infected livestock (Table 4.2) and may contaminate water sources and, thus, may be transmitted and infect humans.

**Table 4.2. Percentage of animals shedding selected zoonotic pathogens**

| Pathogens | % of animals shedding pathogens (no. of pathogens/kg wet weight) | | | | | | |
|---|---|---|---|---|---|---|---|
| | Cattle | Calves | Sheep | Pig | Poultry | Rodents | Waterfowl |
| *Cryptosporidium* sp. | | 20 – 90 ($10^6$ - $10^7$) | 8 - 40 | 5 - 20 | 9 ($10^6$) | 30 | 13 - 100 |
| *Giardia* sp. | | 57 – 97 ($10^6$ - $10^7$) | | | | 10 - 95 | 6 - 50 |
| *Campylobacter* spp. | | | | | | 1 - 10 | 1 - 10 |
| *Salmonella* spp. | 13 | | 4 - 15 | 7 - 22 | | | |
| *Yersinia* sp. | | | | 1 - 10 | | | |
| Pathogenic *E.coli* | 3.5 | | 2.0 | 1.5 - 9 | | | |

Source: Erlandsen, 1994; Geldreich, 1996; Casemore *et al.*, 1997; Medema, 1999; Schijven and Rijs, 2000.

Wild animals are another source of faecal contamination. In general mammals and birds (waterfowl) may shed human pathogens. *Cryptosporidium parvum* has been detected in a wide variety of wild mammals such as foxes, rabbits and a variety of rodents (squirrels, rats, mice, voles, hamsters) (Fayer *et al.*, 1997). Cross transmission has been demonstrated between a number of these mammalian hosts (Fayer *et al.*, 1990). Recent reports indicate that waterfowl may shed viable oocysts of *C. parvum* after ingestion of these oocysts (Grazcyk *et al.*, 1996). Moreover, naturally infected Canada geese were shown to carry and shed the zoonotic strain of *C. parvum* (Grazcyk *et al.*, 1998). Hence, birds that feed on sewage sludge or agricultural lands may ingest *C. parvum* oocysts and are a potential source of water contamination and zoonotic transmission.

115

Several waterborne outbreaks of giardiasis have been related to contamination of water by beavers and also by muskrats, another aquatic mammal with an even higher prevalence of *Giardia* sp. (Moore *et al.*, 1969; Dykes *et al.*, 1980). These reports, however, have been criticised, as the evidence provided was only circumstantial (Woo, 1984; Erlandsen, 1994). Other pathogens that have been implicated in waterborne illness, which also originate from wildlife are *Campylobacter* sp., *Yersinia* sp. and *Salmonella* spp. The carrier incidence of *Salmonella* spp. in waterfowl is generally 1-5%, but may be as high as over 20% in seagulls scavenging near sewage outfalls (Fenlon, 1981). *Campylobacter* spp. has been isolated from birds and rodents (Table 4.2).

In well-protected surface water catchments, upland reservoirs and mountain streams, wildlife may be the most important source of faecal pollution. For example, several cases in the Netherlands show that these systems are most at risk during late winter/early spring, when bird loads on the (partly frozen) reservoirs are high. When thaw sets in, the bird faeces that have collected on the ice enter the water, leading to a peak of contamination with *Campylobacter* or *Cryptosporidium* sp. and *Giardia* sp. (Medema *et al.*, 2000a). In another study, Medema (1999) estimated that waterfowl contributed between 1 and 16% to the *Cryptosporidium* sp. concentration in reservoir water and 4 - 67% to the *Giardia* sp. concentration. However, as these (oo)cysts may not be pathogenic to humans, the significance of this source is a matter of debate.

### 4.2.2    Sources of groundwater pollution

Many practices with domestic wastewater and with livestock manure may lead to contamination of groundwater, these are summarised in Figure 4.1 and outlined in more detail below.

Septic tanks, cesspools, latrines and other on-site systems are widely used for wastewater storage and treatment. The water percolating from these facilities contains viruses, bacteria and parasites and may contaminate groundwater supplies. In the USA, septic tank systems rank highest in terms of the volume of untreated wastewater discharged into the groundwater and they are the most frequently reported source of groundwater contamination (Hagedorn, 1984). Sewers in the unsaturated zone may leak sewage into the soil, and it is likely that the extent of this problem is largely unrecognised. In the saturated zone, sewer breaks will result in groundwater contamination. During heavy rainfall, stormwater collection in sewers may increase the leak rate, leading to increased contamination.

116

There are several types of land application of waste or stormwater, including infiltration, overland flow, wetlands and subsurface injection. Several studies have shown that viruses can be found in the groundwater up to 30m beneath land application sites and can travel several hundreds of meters laterally from the application point (Keswick, 1984). In one case, viruses were demonstrated after heavy rainfall in a sampling site that was previously considered uncontaminated (Wellings *et al.*, 1974).

Stormwater collected in sewers that also transport domestic wastewater can present a major problem. Other than direct discharges to water bodies (which clearly lead to contamination), it may also be disposed of by collection in basins and subsequent drainage to soil. This percolation may transfer pathogens to groundwater, as illustrated by Vaugh *et al.* (1978) who reported viruses in the soil 9 m below a stormwater basin.

Digested or composted sludges from sewage treatment plants are applied to cropland. These sludges contain viruses, parasites and bacteria (Bitton and Farrah 1980; Feachem *et al.*, 1983). Although parasites and many bacterial pathogens are inactivated during thermophilic sludge composting, some viruses survive this treatment (Damgaard-Larsen *et al.,* 1977). In field studies, no viruses could be found in the leachates of sludge disposal sites (Bitton and Farrah, 1980) and it has been suggested that the sludge/soil matrix is effectively retaining the viruses in the sludge. When groundwater tables are high, there may be direct contact between groundwater and sludge and this probably leads to groundwater contamination. The potential risk of contamination of water sources from sewage sludge disposal (along with other routes of transmission) is well recognised and subject to World Health Organization (WHO) guidelines (WHO, 1989; Mara and Cairncross, 1989).

In countries with limited supplies of fresh water, wastewater is used for crop irrigation, either by spray irrigation, overland flow or subsurface infiltration. Wastewater irrigation is also subject to WHO guidelines (WHO, 1989). WHO guidelines for wastewater and excreta use in agriculture are currently undergoing revision.

Many farmers have cellars, tanks or landfills to store manure. Water leaching from these storage sites may contaminate groundwater, especially during periods of rainfall. Storage does reduce the concentration of bacterial pathogens, but (oo)cysts of *Cryptosporidium* sp. and *Giardia* sp. can survive for months in manure (Robertson *et al.*, 1992). The application of animal manure to agricultural lands as fertiliser is common practice throughout the world. Application may be by droppings of animals grazing the land, by spraying a manure slurry over the land or by ploughing or injecting manure into the top layers of the soil. The zoonotic microorganisms present in the manure may leach into the groundwater.

**Figure 4.1. Pathogen transport in an unconfined aquifer**

(From Keswick and Gerba 1980)

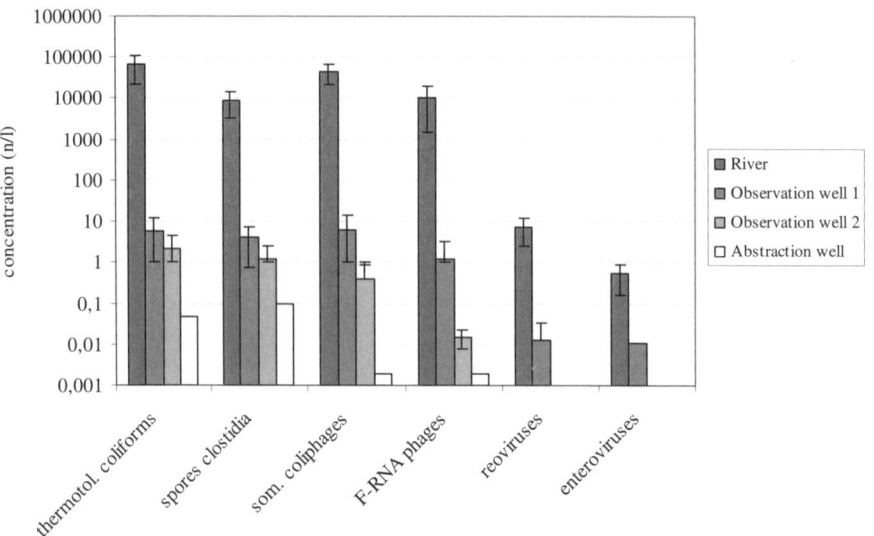

The potential for pathogens from human and animal wastes that are present in the vicinity of wells to contaminate drinking water need special attention. The well construction itself may promote faecal contamination of the aquifer. As the well punctures all layers in the soil above the aquifer, animal droppings or human wastes that are deposited close to the well may travel with percolating rain water directly into the well if the wellhead is not properly protected. Or they may travel along the well wall or in the material surrounding the well in the drill-hole.

## 4.3    Transport and survival

For most faecal pathogens, water is a transmission vehicle rather than a source of pathogens. Most of the enteric pathogens that are discharged into the environment are not able to multiply and need to survive until they are ingested by a suitable human or animal host. This is especially true for obligate parasites, such as the enteric viruses and protozoa like *Cryptosporidium* sp. and *Giardia* sp. but, in general, this also applies to enteric bacteria: *Campylobacter* spp., *Shigella* spp., *Salmonella* spp., enteropathogenic *E. coli*. There is some evidence that *E. coli* can grow in pristine water in tropical rain forests (Rivera *et al.*, 1988), but it is not clear whether this may also be true for enteropathogenic *E. coli*.

### 4.3.1 *Survival in surface water*

The ability of pathogens to survive in surface water differs (Ref – pathogens in surface water – in preparation). In general, survival is prolonged when water temperature is low. Other factors that influence survival include sunlight intensity and the presence of aquatic microorganisms that may use the pathogens as a food source or produce exo-enzymes that cause pathogen disintegration. Adsorption to particles facilitates survival, for example LaBelle and Gerba (1980) found that the survival of poliovirus 1, which was adsorbed to sediment, increased four-fold in an unpolluted zone and 96-fold in a polluted zone. Table 4.3 outlines the disappearance rate and time for a 50% reduction in concentration of a number of pathogens in surface water, using examples of published data.

Another factor that affects survival of both faecal index/indicator parameters and pathogens in surface water is the ability of many bacteria to enter the viable but non-culturable (VBNC) stage of growth (Colwell and Grimes, 2000). Briefly, when stressed by a physical or chemical factor (*e.g.* loss of nutrients, adverse temperature, chlorine), many of the bacteria examined thus far respond to the stress by undergoing a series of structural and physiological changes that result in a dormant or 'non-culturable' stage of growth. They tend to become smaller, less permeable, refractory to cultivation on culture media normally supportive of their vegetative growth, and some lose their flagella.

**Table 4.3. Disappearance rates and reduction times for selected microorganisms in surface water**

| Microorganism | Disappearance rate (per day) | Time for 50% reduction of concentration (days) |
|---|---|---|
| **Pathogens** | | |
| Parasites | | |
| *Cryptosporidium* sp. | 00057 - 0.046 | 15 - 150 |
| *Giardia* sp. | 0.023 - 0.23 | 3 - 30 |
| Viruses | | |
| Enteroviruses | 0.01 - 0.2 | 3 - 70 |
| Hepatitis A | 0.05 - 0.2 | 3 - 14 |
| Rotavirus | 0.24 - 0.48 | 1.2 - 2.4 |
| Bacteria | | |
| *Salmonella* spp. | 1 - 7 | 0.1 - 0.67 |
| *Shigella* spp. | 0.7 | 1 |
| *Vibrio cholerae* | * | * |
| **Index parameters** | | |
| *E.coli* | 0.23 - 0.46 | 1.5 - 3 |
| Coliforms | 0.77 | 0.9 |
| Enterococci | 0.17 - 0.77 | 0.9 - 4 |
| F-RNA phages | 0.01 - 0.08 | 29 - 230 |
| Somatic coliphages | 0.6 - 6 | 2 - 20 |
| *Clostridium perfringens* | 0.0023 - 0.011 | 60 - >300 |

*Vibrio cholerae* is environmentally competent and in unfavourable environmental conditions is thought to survive for long periods in water in a non-culturable state (Colwell and Grimes, 2000). *Source:* DeReignier *et al.,* 1989; Geldreich, 1996; Olson, 1996; Medema *et al.,* 1997; Schijven and Hassanisadeh 2000.

### 4.3.2    *Transport in surface water*

Most enteric pathogens have no means of transport (such as motility) in the aquatic environment other than being transported with the water flow. The pathogens can, therefore, be regarded as biological particles that are transported by advection. Many pathogens readily attach to particles in water (Gerba, 1984; Gerba *et al.,* 1978; Wellings *et al.,* 1974) and these particles largely determine the transport characteristics. Sedimentation of planktonic bacteria, viruses and parasites is very slow and probably not significant in determining transport behaviour, but when attached to particles, sedimentation becomes significant.

Sediments may harbour significant numbers of faecal microorganisms. In bottom sediments from bathing beaches, rivers and streams, van Donsel and Geldreich (1971) found thermotolerant coliforms in concentrations that were 100-1 000 fold higher than those in the overlying waters, and viruses levels in sediments are generally 10-fold higher in sediments than in overlying waters. LaLiberte and Grimes (1982) demonstrated extended survival of *E. coli* inoculated into sediment contained in dialysis bags and placed in a freshwater lake. Re-suspension of sediments, therefore, may give rise to high concentrations of faecal pathogens in water. Rainstorms give rise to re-suspension, as do activities like dredging or shipping (dredge fishing), but the effect of these latter activities appears to be local (Grimes, 1980, 1982).

In temperate lakes thermal stratification may occur during summer and winter. This reduces the exchange of water between the upper and lower layers of lake water. In summer, the quality of the water at the bottom slowly deteriorates due to settling. When de-stratification occurs in the autumn, the water from upper and lower layers mix. This process causes settled particles with coliforms to re-enter the water. In one lake, for example, Geldreich *et al.* (1989) reported that the autumn destratification led to a 10-fold increase in the coliform densities for several weeks; from a level consistently below 10/100 ml in the summer to more than 100/100 ml.

Rainstorms not only result in water quality deterioration through run-off, stormwater discharges and so on, but they also increase water flows. This may result in more rapid transport of faecal pathogens from the contamination source to abstraction sites. Under normal flow conditions, 'self-purification' of water occurs by sedimentation, dilution, sunlight inactivation, predation and starvation. But under rapid flow conditions, self-purification becomes much less significant. In lakes and reservoirs, thermal stratification may strongly reduce the residence time of stormwater. An example of this is provided by Lake Burragorang, a reservoir in Australia. This lake has a length of 40 kilometres and, under normal flow conditions, faecal contaminants are removed through self-purification. The counts of faecal index bacteria and *Cryptosporidium* sp. at the dam-intake are low. A two-year drought reduced the water level of the catchment reservoirs to 60% of their maximum capacity. The area experienced heavy rains in August 1998. These rains flushed the lands and urban areas leading to stormwater discharges. Some tertiary sewage treatment systems were also flooded by the rapidly rising river water. This contaminated flow entered Lake Burragorang. The lake was stratified, with warm water at the top and the colder contaminated water sank to the bottom of the lake and the flow rapidly reached the dam. Turbidity, temperature and *Cryptosporidium* sp. data at the dam-intake showed that there was an interchange at the intake of good quality water from the top layers of the reservoir and poor quality water from the bottom layers within periods of a few days (Deere *et al.*, 2000).

### 4.3.3   *Survival in groundwater*

Survival of microorganisms is an important feature for groundwater systems (REF – ground water book). The mechanisms of elimination of pathogens by soil passage are adsorption and inactivation. The inactivation rate is influenced by many factors as illustrated in Table 4.4.

**Table 4.4. Factors that influence the survival of microorganisms in soils and thus affect their ability to reach groundwater systems**

(Adapted from Gerba and Bitton, 1984)

| Factor | Influence |
|---|---|
| Temperature | Long survival at low temperatures, rapid die-off at high temperatures. For some faecally-derived bacteria high temperatures might give rise to growth. |
| Moisture content | Desiccation is detrimental to most microorganisms (spores excepted). An increased rate of reduction will occur in drying soils. This is of most relevance in the unsaturated zone. |
| Sunlight | More rapid die-off at the soil surface due to UV irradiation. |
| pH | Bacteria die-off more rapidly in acid soils (pH 3-5) than in alkaline soils. The pH influences the adsorption of microorganisms to the soil matrix and indirectly influences survival. |
| Microflora | Soil bacteria and fungi may produce exo-enzymes that damage the structure of faecal microorganisms, while amoebae and other microbiota may feed on them. Bacterial survival is shorter in natural soils than in sterilised soils, but for viruses no clear trend is observed. |
| Organic carbon content | The presence of organic carbon increases survival and may give rise to the regrowth of bacteria. |
| Cations | Certain cations have a thermal stabilising effect on viruses and increase virus survival. Cations also enhance virus adsorption to soil and this indirectly increases survival, as viruses appear to survive better in the adsorbed state. |

Pathogen survival in groundwater has been determined in a number of ways, including:

- Suspending laboratory microorganisms in microcosms.

- Sterile groundwater in flasks in the laboratory or under ambient conditions.

- Membrane chambers in flowing groundwater.

- Dialysis tube in groundwater wells.

The disappearance rates in groundwater are lower than in surface water (see Table 4.5). Viruses survive longer than bacteria. No data on the survival of protozoan parasites in groundwater are available yet, but it can be assumed that these pathogens are able to survive longer than the viruses.

**Table 4.5. Example disappearance rates of enteric microorganisms in natural groundwater**

| Microorganism | Disappearance rate (per day) |
|---|---|
| **Viruses** | |
| Hepatitis A virus | 0.10 - 0.33 |
| Poliovirus 1 | 0.013 - 0.77 |
| Coxsackievirus | 0.19 |
| Rotavirus SA11 | 0.36 |
| Coliphage T7 | 0.15 |
| Coliphage f2 | 0.39 - 1.42 |
| MS2 | 0.063 - 0.75 |
| **Bacteria** | |
| *Escherichia coli* | 0.063 - 0.36 |
| Faecal streptococci | 0.03 - 0.24 |
| *Salmonella typhimurium* | 0.13 - 0.22 |
| *Clostridium bifermentans* spores | 0.00 |

Sources: Matthess *et al.*, 1988; Nasser *et al.*, 1992; Blanc and Nasser, 1996; Schijven and Hassanisadeh, 2000.

### 4.3.4   Groundwater transport

The most important factors in the transport of microorganisms through the subsurface are water flow (the driving force) and soil texture. Most of the studies on groundwater transport have focussed on viruses. The transport of viruses through soil is primarily determined by attachment (Schijven and Hassanisadeh, 2000), while virus inactivation is considered to be less/insignificant (Bales *et al.*, 1995, 1997; Pieper *et al.*, 1997; DeBorde *et al.*, 1998, 1999). Many factors affect the adsorption and transport of microorganisms through soil (Table 4.6). The major factor that affects virus adsorption is pH (Schijven and Hassanisadeh, 2000). At higher pH, electrostatic repulsion increases, resulting in a decreased attachment rate and an increased detachment rate. In most aquifers, surface characteristics of the soil are heterogeneous and also viruses with different isoelectric points may be present. Therefore, dependent on pH and thus on the charge of the virus and soil particles, adsorption of some of these viruses may be irreversible, whereas that of others may be reversible. At pH 7 – 8, adsorption will be mainly reversible.

**Table 4.6. Factors affecting transport of enteric pathogens through soil**

(Adapted from Gerba and Bitton, 1984; Schijven and Hassanisadeh, 2000)

| Factor | Influence on transport |
| --- | --- |
| Soil texture | Fine-textured soils retain viruses, bacteria and protozoa more effectively due to increased interaction and adsorption. Fractured soils, however, are poor retainers of microorganisms. |
| Water flow | Water flow is the driving force of transport and pathogen transport velocities appear to be proportional to the water flow. Increased water flow may remobilize adsorbed microorganisms. |
| PH | Adsorption generally increases when pH decreases, due to reduced electrostatic repulsion. |
| Cations | The presence of multivalent cations ($Ca^{2+}$, $Mg^{2+}$) increases adsorption due to the formation of salt bridges between negatively charged microorganisms and soil particles. |
| Metal hydroxides | Iron hydroxides improve the adsorption of microorganisms. |
| Soluble organics | These can influence transport in various ways: they may compete with microorganisms for attachment sites (humic and fulvic acids compete with viruses), but they may also give rise to microbial activity that enhances attachment and inactivation. |
| Microorganism characteristics | Bacteria and parasites are more readily removed than viruses because of their size (1 - 20 μm versus 20 - 80 nm). Differences in isoelectric points and surface composition determine the adsorption rates. |
| Saturated versus unsaturated flow | Under unsaturated flow conditions, water fills only the small pores. This increases soil-microorganism contact and adsorption. |

## 4.4    Catchment surveys and catchment protection

As indicated earlier, catchment protection is an essential step in safeguarding the microbial quality of drinking water. The basic concept of catchment protection is to know the catchment hydrology/hydrogeology and sources of pathogen contamination in the catchment in order to:

- Select the most appropriate site for drinking water abstraction or well placement.

- Be able to select appropriate catchment monitoring and/or protection measures.

- Predict the occurrence of peak events.

A sanitary survey of the catchment area can identify sources of faecal contamination (sewage treatment plants, sewer overflows, agricultural areas with manure storage or land deposition, high waterfowl numbers etc.). It can also identify if certain climatological (heavy rainfall), environmental (high animal loads) or man-made conditions (agricultural practices, tourism) are likely to give rise to peak contamination events of the source water.

A survey of sources can be done without the use of index parameters by mapping the catchment and the sources of faecal contamination present. This is the basic hazard assessment step in the catchment. In a second stage, a survey can be conducted with the well-established microbial indices of faecal pollution (*E. coli*, enterococci, spores of *Clostridium perfringens*), non-microbial water quality data (*e.g.* turbidity, temperature, pH, conductivity) and hydrologic data (measurements of flow and precipitation). This will provide more accurate and quantitative information about the quality of the waters in the catchment areas and the effect of transport on the level of faecal pollution. If the methods and resources are available, the inclusion of F-RNA bacteriophages in such a survey is likely to improve its prediction with regard to virus hazards.

Other, non-microbial parameters have been suggested as indicators of contamination with domestic wastewater. These are compounds that are used in the household such as boron (used as whitener in washing powders) and caffeine, and other human excretory products such as secretory IgA, sterols and urobilin. None of these, has been demonstrated to be widely applicable, but may be useful for specific purposes.

### 4.4.1 Surface water

#### 4.4.1.1 Catchment survey

For surface water systems, this inventory should include:

- Location and size of discharges of treated sewage.

- Location and size of discharges of untreated sewage.

- Location and size of sewer overflows and conditions initiating overflow.

- Location of sewage sludge deposits and ability of contaminants to enter surface water.

- Location, type, frequency, conditions and size/weight of manure application on land in agricultural areas.

- Location and type of manure deposits and ability of contaminants to travel to surface water.

- Presence of high numbers of wild mammals and birds on or around the surface water.

For all these aspects, special attention should be given to circumstances that may lead to peak contamination events. A simple classification relating to the estimated significance of the pollution source is helpful and can be based on its nature, size, the transport time and its distance from the source water.

The catchment survey should result in:

- An inventory of the contamination sources.

- A classification of their significance.

- An inventory of conditions that may give rise to peak contamination events.

During the survey possible risk management measures may be identified and, after the survey is completed, it is generally possible to classify catchment protection measures according to their (estimated) impact on the improvement of the source water quality or prevention of peak events.

A basic catchment survey is not sophisticated (although geographical information system (GIS) techniques are helpful). Despite this, however, they are not very well established within the water community. There are some examples of catchment surveys, such as those in several German reservoir catchments (Feuerpfeil and Bischoff, 2001), but in general, emphasis has more commonly been placed on treatment. This may be because treatment is totally within the control of the water supplier, while catchment protection may involve many different bodies and interest groups and is, therefore, more difficult to manage. However, the recognition of *Cryptosporidium* sp. as an important waterborne pathogen and its resistance to disinfection has boosted interest in catchment protection, especially in North America. Following an incident in Australia in 1998, where water in distribution was found to be contaminated with *Cryptosporidium* (probably as a result of high levels within the catchment following heavy rainfall - McLellan, 1998) a Catchment Authority was set up, with their first step being to make an inventory of contamination sources (Deere *et al.*, 2000).

### 4.4.1.2    *The use of microbial parameters as an index of faecal pollution*

An initial catchment survey focussing on the identification of sources can be conducted without the use of microbial indices, however, such measurements may help to determine the significance of any identified pollution sources and also the behaviour and transport of faecal contamination in surface water. They are especially helpful in assessing the occurrence of peak events and may even be used to predict these. The principal microorganism for this purpose is *E. coli*, as it is present in all faecal contamination of concern and the assay is inexpensive, simple and widely used. Its presence in surface water indicates recent faecal contamination and therefore a potential health hazard. Thermotolerant coliforms can be a suitable alternative, but where these are used attention must be given to the possible presence of waste effluents with a high carbohydrate content, as they may harbour *Klebsiella* sp.

As *E. coli* is not as environmentally long-lived as many pathogens (*i.e.* viruses and protozoa) it is most useful in identifying recent contamination. Additional, complementary tests examining for the more robust enterococci and the spores of *Clostridium perfringens* can shed light on less recent faecal contamination.

Due to high thermophilic background growth on the culture media and the potential multiplication of thermotolerant coliforms (and even *E. coli*) in the environment (Rivera *et al.*, 1988; Byappanahalli and Fujioka, 1998), the use of some microbial parameters to assess tropical source water quality is problematic. *Clostridium perfringens* spores appear to be the most appropriate parameter for assessing faecal contamination in tropical climates (Fujioka, 2001).

*Case study: Using monitoring within a water safety plan to prevent contamination*

One of the catchments within the Melbourne Water area in Australia is a protected natural mountain area with no human settlements. The water is held in reservoirs and is only chlorinated before distribution to the consumers. Catchment surveys and monitoring showed that rainfall led to deterioration in the water quality in the tributaries to the reservoir with both relatively high turbidity and concentrations of faecally derived bacteria. To attempt to remedy this, a standard operating procedure was set up, using flow diversion as a control measure to prevent water of relatively poor quality entering the reservoir (Deere *et al.*, 2000).

### 4.4.1.3   *The use of pathogenic microorganisms*

It is generally considered that surveys assessing pathogenic micro-organisms give the most direct information about their sources. There are, however, many different pathogens, they are present in relatively low concentrations and require large sample volumes and pathogen detection methods (if available) are generally high-tech, time-consuming and expensive. Thus pathogen surveys rarely allow large numbers of samples to be taken. Such surveys are, therefore, best preceded by catchment surveys, sanitary inspection and faecal index investigation and then targeted to specific research questions, as illustrated by the following case study.

*Case study: First flush*

In the California State Water project in the USA, several lakes are used as source waters. Two of these lakes are Castaic Lake in the Castaic Mountains and Silverwood Lake in the San Bernardino National Forest. No agricultural, industrial or sewage discharges were identified in the streams sampled during sanitary surveys, but livestock and wildlife were present in the vicinity of the creeks. For the *Cryptosporidium* and *Giardia* monitoring programme, sampling sites were selected in the lakes and in tributaries and three sampling strategies were used:

- Large volume (100 l) sampling with filters.
- Unattended stormwater grab samplers (5 l).
- 4 l grab samples for storm events other than first flush.

The first flush during storm events contained very high numbers of *Cryptosporidium* sp. and *Giardia* sp. (Table 4.7). These high concentrations go unnoticed in the filter sampling that was not driven by precipitation but also the grab samples during storm events did not detect these high numbers (Stewart *et al.*, 1997).

**Table 4.7. Results of the Castaic and Silverwood lake surveys with different monitoring strategies**

| Sampling strategy | *Cryptosporidium* sp. | | *Giardia* sp. | |
|---|---|---|---|---|
| | % positive | Range (no/100 l) | % positive | Range (no/100 l) |
| Filter | 10 | 3 - 415 | 29 | 2 - 119 |
| First flush | 35 | 46 - 41 666 | 60 | 25 - 16 666 |
| Grab sampling | 19 | 3.4 - 647 | 19 | 42 - 2428 |

Hence, the first flush during storm events carry the highest load of *Cryptosporidium* and *Giardia* to the streams and should be included in monitoring programmes or somehow accounted for. The first flush samplers were inexpensive, but were vulnerable to theft and vandalism.

## 4.4.2    Groundwater

### 4.4.2.1   Catchment survey

As outlined earlier, primary sources of groundwater contamination are human and animal wastes and the infiltration of faecally contaminated surface water (including rivers). As groundwater treatment generally does not include major barriers against pathogens, groundwater catchment protection acts as the principal barrier. However, it should not necessarily be assumed that groundwater does not require filtration to remove protozoa (*e.g.* where the filtration provided by the ground between a river and a well is inadequate to remove protozoa) or disinfection to inactivate viruses.

The catchment survey needs to collect information about the sources of contamination in the catchment, the hydrogeology of the groundwater system and the high risk factors such as the possibility of rapid pathogen transport through pores/fractures or contamination of wells, drains or adits from nearby sources.

*Geohydrological survey*

This should include investigation of the following:

- Catchment area (if no detailed information is available, an estimation based on abstraction rate and thickness and porosity of the aquifer is appropriate).

- Soil layer texture and composition; the presence, thickness and integrity of confining soil layers.

- Flow lines of water to the abstraction sites in different layers.

- Presence of solutions, thin unsaturated zones, surface-water aquifer contact.

*Survey of the sources of contamination and high risk factors*

Sources of contamination could include, sites for disposal of sewage, treated sewage or sewage sludge, sites for disposal or land application of manure, areas with irrigation with treated or untreated domestic wastewater, septic tanks, cesspools, latrines, waste dumps, manure storage facilities and so on (see also REF groundwater book).

High risk factors are those that result in the rapid transport of water to the groundwater source, such as heavy rainfall or infiltration of river water.

*Well-head, well, borehole protection and protection of shallow aquifers and drains*

The aspect should include an assessment of sources of contamination in the vicinity of the wells (grazing animals, septic systems, sewers and so on), the well-head integrity and the integrity of the soil around the well (placement and integrity of well sheets, clay layer or concrete slabs around the well). An example inspection form from the WHO is shown in Figure 4.2 and Box 4.1 (WHO, 1997). For shallow aquifers and drains, areas of surface-aquifer contact or small unsaturated zones (especially during rainfall) should be examined, as should the integrity of ventilation shafts and manholes, and also the procedures for opening and entering manholes.

## Box 4.1. Example survey form for tubewell with hand-pump

### (WHO, 1997)

I   Type of facility   TUBEWELL WITH HAND-PUMP

  1.  General information:  Health centre ...................................................................

                                Village ...........................................................................

  2.  Code no.—Address ........................................................................................

  3.  Water authority/community representative signature .........................

  4.  Date of visit ......................................

  5.  Water sample taken? ....... Sample no. ......... Thermotolerant coliform grade .........

II   Specific diagnostic information for assessment                                   Risk

  1.  Is there a latrine within 10 m of the hand-pump?                               Y/N

  2.  Is the nearest latrine on higher ground than the hand-pump?                   Y/N

  3.  Is there any other source of pollution (e.g. animal excreta, rubbish,          Y/N
      surface water) within 10 m of the hand-pump?

  4.  Is the drainage poor, causing stagnant water within 2 m of the
      hand-pump?                                                                     Y/N

  5.  Is the hand-pump drainage channel faulty? Is it broken, permitting
      ponding? Does it need cleaning?                                                Y/N

  6.  Is the fencing around the hand-pump inadequate, allowing animals in?           Y/N

  7.  Is the concrete floor less than 1 m wide all around the hand-pump?             Y/N

  8.  Is there any ponding on the concrete floor around the hand-pump?               Y/N

  9.  Are there any cracks in the concrete floor around the hand-pump which          Y/N
      could permit water to enter the well?

10.  Is the hand-pump loose at the point of attachment to the base so that           Y/N
      water could enter the casing?

                                         Total score of risks .................... /10

Contamination risk score: 9–10 = very high; 6–8 = high; 3–5 = intermediate;
                          0–2 = low

III   Results and recommendations

The following important points of risk were noted: ............................... (list nos 1–10)
and the authority advised on remedial action.

Signature of sanitarian .........................................

**Figure 4.2. Diagram to accompany example inspection form (Box 4.1) for tubewell with hand-pump**

(WHO, 1997)

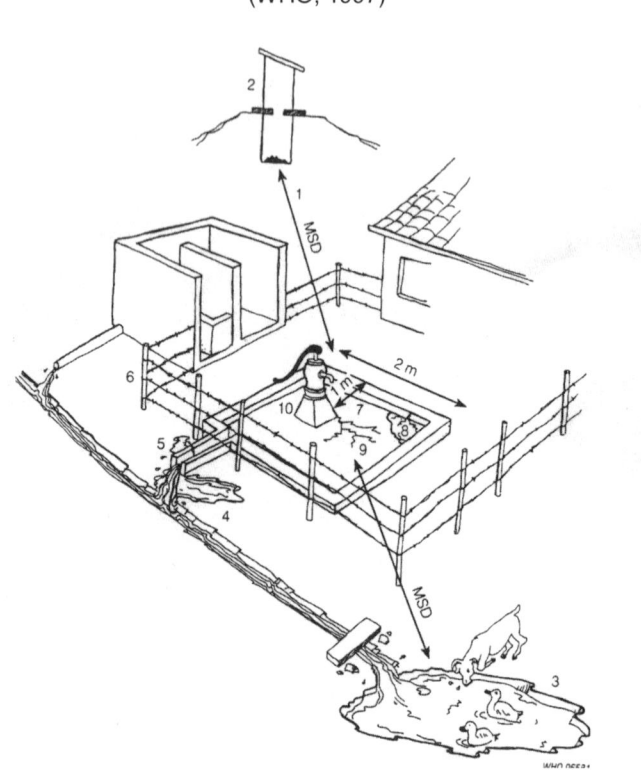

MSD: minimum safe distance determined locally.
1-10 refer to the survey form (Section II) – Box 4.1.

In some countries, inventories of sources of groundwater contamination (both for microbial and chemical contaminants) are better established than in surface water systems. In these countries, groundwater protection regulation has been issued that relies on the division of the catchment area into zones of different vulnerability to contamination. The zones are defined by (average) transit time of the water from the land surface to the source water. Activities that may lead to groundwater contamination are restricted in these protection zones.

An outbreak of cryptosporidiosis in North London (UK) was subsequently traced to a groundwater supply that was vulnerable to infiltration of surface water containing *Cryptosporidium* sp. (DWI, 1998). This and several other incidents with groundwater supplies triggered the UK Group of Experts to recommend that groundwater systems should be evaluated for potential contamination risk (Bouchier, 1998). The groundwater protection practice at the time was based (as in many countries) on land surface zoning according to travel times of the water from the land surface to the groundwater sources and the restriction of contaminating activities in the most vulnerable zones. Bouchier (1998) evaluated these practices for their protective value against microbial pollution, especially with *Cryptosporidium* sp. It was concluded that this approach formed a sound basis for assessing vulnerable groundwater supplies, but also that there were some limitations. An important limitation is that by-pass features, which allow rapid transport of water with contaminants to groundwater, were not incorporated in the vulnerability assessment. By-pass flow may occur in many of the British carbonate aquifers. Similarly, surface water-aquifer interactions that may occur in valley-bottoms (surface water recharge) and upper catchments were not incorporated in the vulnerability assessment. Bouchier (1998) recommended the inclusion of an additional vulnerability class in the zoning scheme. This extreme vulnerability class would apply to areas with the combination of contaminated surface water and rapid access points (solution features, sinkholes, karst or pseudo karst features, mines and aggregate extraction sites).

The need for the inclusion of rapid access of surface water to groundwater as an important factor in vulnerability assessment was illustrated by the fact that eight of the nine suspected cases of groundwater contamination with *Cryptosporidium* sp. in the UK were associated with adited wells, collectors, spring galleries and former mines with adits (Morris and Foster, 2000). Groundwater supplies in rural settings were more commonly affected than sites in urban settings. Fissure flow, dual porosity flow and intergranular flow were all represented in these cases. Intergranular flow appeared only to be important in settings where the residence time in the aquifer was very short, such as in river gravels close to a surface water course.

The Expert Group listed the factors of a groundwater system that need to be considered for assessing the risk of contamination with *Cryptosporidium* sp. (Table 4.8) and gave guidance on techniques to determine/verify the significance of these factors. Simple qualitative ranking may help to prioritise the different hazards, however, Morris and Foster (2000) stress the need to focus on the individual water supply when applying ranking, appreciating the unique hydro-geological, operational and contamination sources setting of each supply system.

**Table 4.8. Factors for consideration in the risk assessment of groundwater contamination**

(From Bouchier, 1998)

| Predisposing groundwater to *Cryptosporidium* sp. risk | Possible verification techniques |
|---|---|
| **Catchment factors** | |
| High wastewater returns, including sewage effluent to river reaches, especially under baseflow conditions | Hydrochemistry, microbiology, hydrometry |
| Livestock rearing in inner catchment, especially if intensive | Farm survey |
| Likely *Cryptosporidium* sp. - generating activities in catchment – *e.g.* abattoirs | Economic activity survey |
| Urbanising catchment | Land registry survey |
| Livestock grazed or housed near wellhead | Site inspection |
| **Hydrogeological factors** | |
| Known or suspected river aquifer connection nearby | Flow gauging, modelling, hydrochemistry |
| Unconfined conditions with shallow water table | Well-water level monitoring |
| Karst or known rapid macro-fissure flow conditions, especially in shallow groundwater | Field mapping, farm survey |
| Patchy drift cover associated with highly contrasting aquifer intrinsic vulnerabilities | Field mapping, shallow drilling |
| Solution features observed or inferred in catchment | Field mapping |
| Shallow flow cycles to springs | Tracing, hydrochemistry, water temperature logging |
| Fissure-dominant flow (as suggested by high transmissivity or specific capacity) | Downhole fluid/flow logging, pumping test analysis |
| **Well/raw water source factors** | |
| Supply source tapping shallow flow systems (*e.g.* adits, springs, mine galleries) | Check site plans, tracing |
| Adits with upbores or construction-stage ventilation shafts | Check site plans, site inspection |
| Poor casing integrity | CCTV, geophysical logging |
| Masonry linings above pumping water level without additional sanitary seal | CCTV, check site plans |
| Sewer/septic tank/slurry pit systems near wellhead or above adits | Site inspection |
| Inadequately fenced source especially around spring boxes, catchpits, galleries | Site inspection |
| Old, poorly documented well construction | Site plans/BGS national well record archive |

CCTV: Closed circuit television. BGS: British Geological Survey.

*4.4.2.2   The use of microbial indicators as an index of faecal pollution*

Groundwater is not easily accessible for monitoring faecal contamination. Sources of faecal contamination that have been identified in the sanitary survey can be monitored and may give quantitative information on the size of the sources but not necessarily on their impact on groundwater quality.

The subsurface catchment can be monitored for faecal contamination with observation wells, if the hydro-geology has been well established. This is especially true for karst or fractured bedrock aquifers, because the water flow is difficult to establish. The ideal placement of observation wells is between the sources of contamination and the abstraction wells. Detailed guidance on sampling of groundwater and well drilling and placement can be found in McNabb and Mallard (1984). Monitoring of the observation wells for index parameters can give information about the efficiency of soil passage. Microbial parameters that can give information about the presence of faecal contamination are *E. coli,* bacteriophages and spores of *Clostridium perfringens.* The application of these parameters in observation wells is similar to source water monitoring for groundwater systems. Important recommendations are:

- Sample volumes should be as large as achievable (Fujioka and Yoneyama, 2001 recommend analysis of 1 000 ml samples).

- Sampling strategy should include regular sampling and high risk event sampling. In this respect, observation wells should be placed in the flow lines from high risk areas to source water and preferably leave sufficient residence time to take appropriate actions (*e.g.* shut-down of wells).

- When resources/available methods are limited, use *E. coli* testing. When resources are available include testing for enterococci, spores of *Clostridium perfringens* and F-specific RNA phages if sewage or other human wastes are thought to be important pollution sources.

*Case study: The importance of large volumes*

To increase production capacity, Water Company Limburg (in the Netherlands) has constructed additional wells along the banks of the River Meuse. Hydrological calculations indicated that 10-15% of the abstracted water is bank filtrate. The aquifer consists of coarse and fine gravel with sand and the average residence time of the river water in the soil is 0.5–3 years. One of the wells close to the river (150 m) extracts 50% river water and the estimated residence time to this well is 45-65 days. Due to the nature of the Meuse (rain-fed river), sudden sharp rises in the river level occur that can lead to infiltration of previously unsaturated soil layers with high velocities and short residence times. To determine the relationship between the residence time (or distance of soil passage through the aquifer) and the removal of pathogenic microorganisms, large volume samples (up to 100 l) were taken from the river and from two observation wells located between the river and the pumping well closest to the river. The samples were analysed for the presence of thermotolerant coliforms, spores of sulphite reducing clostridia, *Cryptosporidium* sp., *Giardia* sp., entero- and reoviruses, somatic coliphages and F-specific RNA phages. All were present in the River Meuse throughout the study period. In the observation well at a distance of 8 m from the river, thermotolerant coliforms, spores of sulphite-reducing clostridia, bacteriophages and reoviruses were detected (Figure 4.3). No enteroviruses were found. *Cryptosporidium* and *Giardia* were tested once and not detected. In the observation well further from the river only thermotolerant coliforms, spores and bacteriophages were detected. In the pumping well, thermotolerant coliforms, spores of sulphite-reducing clostridia and somatic coliphages were occasionally detected. The highest concentrations in the well were observed after the water level in the River Meuse had been high and the distance and travel time to the observation well was relatively short.

The first metres of soil passage resulted in relatively efficient removal, probably because this includes passage through the relatively impermeable layer of river sediment on the bank. Strict hygiene measures were necessary to prevent contamination of the observation wells from the soil surface. The occasional presence of several indices of faecal pollution in the well indicated that microorganisms (including pathogens) from the river may pass through the soil into the well, especially during high river flows. The results were used to design an abstraction programme in which the well(s) close to the river are shut down in case of a sharp rise in river flow (Medema *et al.*, 2000b).

**Figure 4.3. Removal of microorganisms by bank filtration in a gravel-sand soil**

(Medema *et al.*, 2000b)

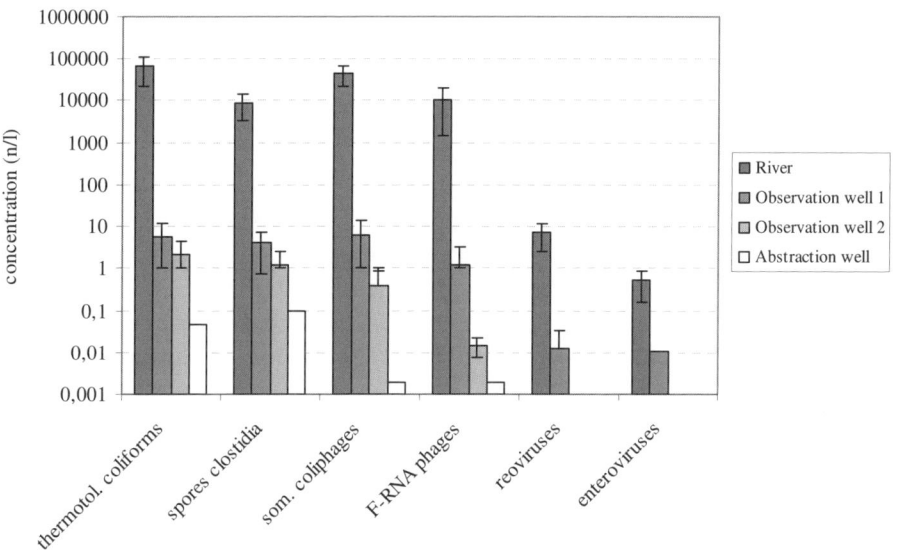

## 4.5    Source water quality

For source water quality assessment, the different nature of the sources means that a clear distinction between surface water and groundwater sources is necessary. For surface water, assessment of the microbiological quality of the source water is essential in both the design and the operation phase of drinking water treatment:

- To design an appropriate treatment system that transforms the source water into safe drinking water.

- To evaluate if an existing treatment system is able to provide safe drinking water.

- To target the treatment to cope with variations and peak events in the faecal contamination of source water.

In most groundwater systems, faecal contamination is generally low or even absent. If the groundwater is influenced by surface water then the monitoring programme should also include the surface water (this can be targeted to conditions or times when surface water influence is likely). For the most part, monitoring is used to verify the purity of the groundwater, rather than to determine treatment goals, however, if the groundwater is influenced by faecal contamination, then monitoring has both functions (*i.e.* verification of the efficacy of the soil passage, and assessment of the additional treatment level required).

The first stage of the assessment of source water quality is monitoring for the traditional microbial indices of faecal pollution, possibly supplemented with bacteriophages. This monitoring programme should be frequent in order to identify short-term variations in water quality. If certain climatological, natural conditions or human practices lead to an increase in the faecal contamination of the source water, monitoring intensity (frequency and potentially the number of parameters) should be increased during the period when these conditions occur. Monitoring for pathogens is (as for catchment surveys) secondary to index monitoring and should only be considered when there is a strong suspicion of contamination. Although it should be noted that pathogen monitoring can provide useful information. In source water, for example, quantitative information on pathogen occurrence provides data that can be helpful in setting treatment goals. In groundwater supplies it can be useful as a way to determine the ability of pathogens to travel through the soil. However, when only low levels of index parameters are detected, the probability of finding pathogenic microorganisms decreases.

### 4.5.1    Surface water

#### 4.5.1.1    The use of microbial parameters to set treatment goals

Microbial parameters can be used to determine the level of faecal contamination at a specific surface water site. The more contaminated a surface water source, the more treatment is necessary to produce safe drinking water. This is the basis of microbial guidelines for raw water quality, such as those in the EC directive 75/440/EEC (Table 4.9).

**Table 4.9. Thermotolerant coliform standards for surface water intended for the abstraction of drinking water**

(EC, 1975)

| Category | Thermotolerant coliform concentration/100 ml | Treatment requirement |
|----------|-----------------------------------------------|------------------------|
| A1 | 20 | Simple physical treatment and disinfection |
| A2 | 2 000 | Normal physical treatment and disinfection |
| A3 | 20 000 | Intensive physical treatment and disinfection |
| A4 | >20 000 | Not suitable for drinking water production |

Many studies have looked at the relationship between microbial indices of faecal pollution and pathogens in surface water. Although this has been subject of much debate, the consensus appears to be that there is a general, coarse relationship between the indices of faecal pollution and pathogen concentrations. Basically, sewage contains higher concentrations of both the index parameters and pathogens than treated sewage and this is again higher than in surface water and after reservoir storage. So, based on the microbial index parameter concentration, a rough estimate of pathogen concentrations can be given (Payment et al., 2000). When it comes to predictions of the pathogen concentration at a particular site, however, the correlation found in general (and even at the location itself) is generally too uncertain to be able to predict the pathogen concentration with less than one to two log-units uncertainty margin on either side of the estimate (Havelaar, 1996).

*4.5.1.2 The use of pathogens to set treatment goals*

LeChevallier and Norton (1995) sampled raw waters supplying 72 drinking water plants in the USA for the occurrence of *Cryptosporidium* sp. and *Giardia* sp. They calculated the required treatment efficiency on the basis of the concentration of parasites in the raw water and the maximum concentrations allowable in drinking water (determined by the $10^{-4}$ risk of infection value used in risk assessment studies in the USA and elsewhere – see Section 1.5.1). Similar studies were conducted in Canada (Payment et al., 2000; Barbeau et al., 2001) and in the Netherlands (Medema *et al.,* 2000c), however, current detection methods make interpretation of such monitoring data for *Cryptosporidium* sp. and *Giardia* sp. difficult. The recovery efficiency of the methods are low and variable and the methods do not discriminate between live and dead (oo)cysts nor do they identify which (oo)cyst types are infective to

139

humans. The conservative approach is to consider all (oo)cysts to be infective, but this may give rise to higher investments in treatment processes than necessary to protect public health adequately. The recent developments in molecular detection methods may bring the required specificity within reach but will not resolve the problems associated with recovery.

Virus surveys have been conducted for the same purpose. The occurrence of enteroviruses in raw water has been monitored in many countries and levels have been found to range from 0.1-100 /l (Block *et al.,* 1978; Nestor *et al.,* 1981; Payment, 1981; Lucena *et al.,* 1982; van Olphen *et. al.,* 1984). While many of the enteric viruses can be cultivated and concentrated from water samples with reasonable efficiency, some of the viruses responsible for waterborne outbreaks cannot be cultivated (*e.g.* Norwalk, hepatitis E and so on). If the objective is to estimate a general level of viral contamination, current methods provide a fair assessment. In the case of outbreak investigation, molecular methods can be used to supplement current methods by providing tools to identify the aetiological agent (see Chapter 7). A combination of cell culture and molecular methodology is being developed (see Chapter 8) which may provide good data for source waters.

### 4.5.1.3   Peak events

Precipitation events can lead to a high pathogen load in the source water. Although this is generally true for all faecal pathogens from domestic and agricultural sources, recent research has focussed primarily on *Cryptosporidium* sp. and *Giardia* sp. This research can be used to illustrate the significance of peak events and the strategies to monitor for peak events such as storms (Gibson *et al.,* 1998; Stewart *et al.,* 1997). Several authors have found a relationship between rainfall and high concentrations of *Cryptosporidium* sp. and *Giardia* sp. (Poulton *et al.,* 1991; Hansen and Ongerth, 1991; Atherholt *et al.,* 1998). The high concentrations were associated with agricultural run-off, re-suspension of river sediments and sewer overflows.

Rainfall also leads to increased water flows and may result in the short-circuiting of pre-treatment reservoirs, as outlined earlier. Flooding, as a result of exceptional rainfall events, can lead to even more extreme contamination. As the floodwater can wash the contents of complete sewage systems, sewage treatment works and sludge disposal stores into surface water and may lead to power failure, mains break and even submersion of drinking water treatment facilities.

*Case study: The importance of frequent and/or event-based monitoring*

The Delaware River flows through the States of New York, Pennsylvania and New Jersey (USA). 1.2 million people live in the river basin. Sources of faecal pollution include combined storm and domestic wastewater sewers, septic systems, discharges of treated domestic wastewater, water recreation and run-off. The Trenton Water Works collects water from this river for the production of drinking water by flocculation, alum coagulation, sedimentation, rapid sand filtration and chlorine disinfection. The abstracted water was sampled monthly for *Cryptosporidium* sp., *Giardia sp.*, indicator bacteria, coliphages and other parameters (*i.e.* turbidity, particles, suspended solids, temperature, river flow). Additionally, the sampling frequency was increased to daily samples (Monday – Friday) in three consecutive weeks during the winter and this was repeated in spring, summer and autumn. To determine the effect of this difference in sampling strategy, the monthly samples were compared to all data and are summarised in Table 4.10 (Atherholt *et al.*, 1998).

**Table 4.10. Comparison of sampling strategies: monthly sampling versus all data**

(Atherholt *et al.*, 1998)

|  | *Cryptosporidium* sp. | | *Giardia* sp. | |
|---|---|---|---|---|
|  | Monthly | All | Monthly | All |
| Percent detection (%) | 92 | 88 | 50 | 40 |
| Geometric mean (n / 100 l) | 51 | 51 | 21 | 24 |
| 90-percentile (n / 100 l) | 134 | 160 | 20 | 40 |
| Minimum (n / 100 l) | 20 | 20 | 20 | 20 |
| Maximum (n / 100 l) | 140 | 800 | 40 | 280 |

While the monthly samples did reflect the central tendency for (oo)cyst occurrence, it underestimated both the 90-percentile and maximum occurrence of the parasites. In addition, LeChevallier and Norton (1995) examined the relationship between *Cryptosporidium* sp., *Giardia* sp. and the other water quality parameters. Although no correlation was found to be consistently present in all time series tested, correlations between protozoa concentrations and turbidity, river flow and thermotolerant coliforms (or *E. coli*) were repeatedly found. The authors developed a simple model for the prediction of peak events based on these indicators of pollution.

*Case study: Turbidity and spores of Clostridium perfringens as indices of pollution*

The River Meuse is a river fed primarily by rain water, which flows from Northern France, through eastern Belgium and the Netherlands to the North Sea. The river receives discharges of treated and untreated domestic wastewater and flows through agricultural areas with a high density of livestock. The river water is abstracted near Keizersveer for the production of drinking water by reservoir storage (typically five months in a series of three reservoirs) and coagulation/filtration with chlorination or ozonation and granular activated carbon filtration. In 1994, the river water was sampled weekly for indicator bacteria, *Cryptosporidium* sp. and other parameters such as turbidity and temperature. The Meuse is a typical rain-fed river, with high flows and high turbidities in winter and spring, due to rainfall and melting snow. At this time the level of faecal pollution, as judged by the concentration of index bacteria, is also relatively high and the water temperature is low. Monitoring showed that several, but not all, of the peak concentrations in *Cryptosporidium* sp. coincided with a peak in turbidity (Figure 4.5, week 3,4,5, week 12, week 15, week 44 and week 50) (Medema, 1999). A turbidity peak therefore indicates the potential presence of high *Cryptosporidium* sp. concentrations (a similar relationship has been observed in the Delaware river (USA), Atherholt *et al.,* 1998).

**Figure 4.5. Coincidence of peaks in turbidity measurements with peaks in *Cryptosporidium* sp. counts in river water**

(Medema, 1999)

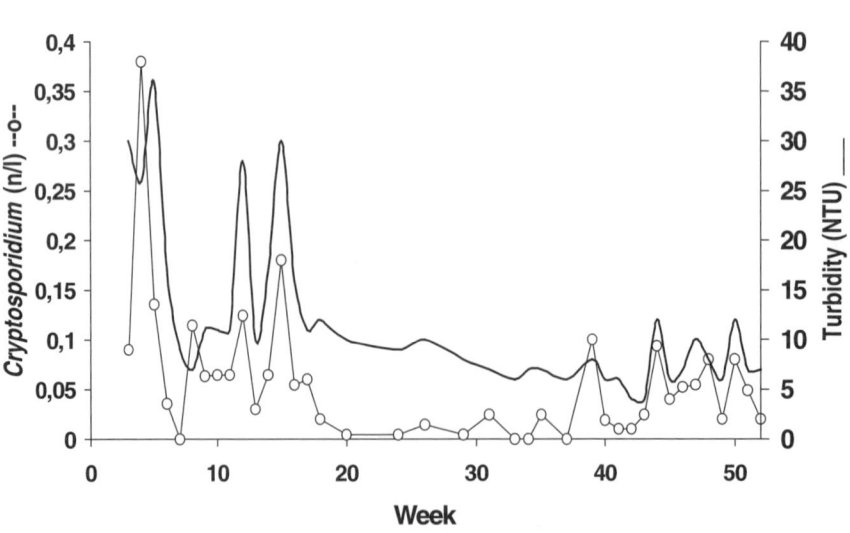

The concentration of *Cryptosporidium* sp. was positively correlated with flow and turbidity, and with faecal indicator bacteria, especially with clostridial spores (Table 4.11). Spores and oocysts are both very persistent in the aquatic environment, and it may be this feature that is the basis of the correlation.

**Table 4.11. Correlation between *Cryptosporidium* sp. concentrations and other water quality parameters in the River Meuse (product-moment correlation coefficients)**

(Medema, 1999)

| Water quality parameter | r-value |
|---|---|
| Spores of sulphite reducing clostridia | 0.75[***] |
| Spores of *Clostridium perfringens* | 0.76[***] |
| Total coliforms | 0.51[**] |
| Thermotolerant coliforms | 0.58[***] |
| Faecal streptococci | 0.57[***] |
| Temperature | -0.44[**] |
| Chlorophyll A | -0.27 |
| River flow | 0.61[***] |
| Turbidity | 0.66[***] |
| $NH_4^+$ | 0.38[*] |

[*]    significant at the 0.05 level.
[**]   significant at the 0.01 level.
[***]  significant at the 0.001 level.

This case study shows that raw water turbidity can be used as an easy and rapid (even on-line) indicator of the presence of *Cryptosporidium* sp. (and probably other pathogens related to faecal contamination). The turbidity measurements serve as a trigger to stop the intake of river water into the reservoirs, while the reservoirs allow the continuation of drinking water production. The study also showed that spores of *Clostridium perfringens* are the best indices of the presence of *Cryptosporidium* sp., probably because both spores and oocysts are robust survival stages.

## 4.5.2    Groundwater

Groundwater supplies are often tested for the same parameters as surface water supplies, including coliform and thermotolerant coliform testing for detection of faecal contamination. As has already been discussed the presence of thermotolerant coliforms indicates the presence of faecal contamination and the potential presence of pathogens. Absence of thermotolerant coliforms, however, does not necessarily ensure that pathogens are absent. Enteric viruses, pathogenic bacteria such as *Yersinia* sp. (Lassen, 1972) and protozoan parasites have been found in groundwater (generally in large volume samples), while tests for *E. coli* (generally small volume samples) were negative.

Differences in attachment rates and survival lead to differences in the ability of microorganisms to travel long distances through the aquifer. In aquifers of fine-textured soil that are vulnerable to faecal contamination from human waste, the risk of penetration of enteric viruses through the soil is higher than for bacteria and parasitic protozoa. This is because of their small size and low attachment rates, and means that bacteriophages are the most suitable parameters for assessing the contamination of the abstraction well. In coarser or fractured soils, bacteria and protozoa can travel long distances and the persistence of protozoa such as *Cryptosporidium* sp. make these pathogens critical for groundwater source quality. Spores of *Clostridium perfringens* have been suggested as an index for these persistent pathogens (as outlined earlier). Estimates based on spores of *C. perfringens* will be conservative, because of their smaller size (1 μm versus 4 μm for oocysts) and their even greater persistence in the environment (Hijnen, unpublished data). The use of spores of sulphite-reducing clostridia as an approximation of *C. perfringens* spores is less appropriate in groundwater systems as (anaerobic) soil is a natural habitat for *Clostridium* spp.

If surface water is a potential source of groundwater contamination, the presence of freshwater algae in well water suggests that biological particles of this size (and thus also protozoan parasites, viruses and bacteria) may pass the soil and contaminate the abstracted water. This was demonstrated in a survey in the USA. Moulton-Hancock *et al.* (2000) showed that the presence of aquatic microbiota was a significant predictor of the presence of *Cryptosporidium* sp. and *Giardia* sp. The microbiota included algae, rotifers, fungi, arthropods, colourless flagellates, nematodes, amoebae and gastrotrichs. It was found to be possible to differentiate between high, moderate and low risk groundwaters on the basis of the quantity and composition of the microbiota present. They suggested the use of a presence/absence test for algae as a simple tool for the prediction of the presence of *Cryptosporidium* sp. and *Giardia* sp.

The sample volumes that are used for groundwater monitoring are usually low (100 ml, despite recommendations to use larger volumes – Fujioka and Yoneyama, 2001) and the sampling frequency is generally not high (weekly - monthly). The application of larger volumes and/or increased frequency and sampling of individual wells/adits increases the sensitivity of tracing faecal contamination. The monitoring programme should not only cover regular sampling, but also be designed to examine high risk factors, such as rainfall or the presence of vulnerable sites within the groundwater abstraction system.

### 4.5.2.1 *Turbidity and temperature profile*

Non-microbial parameters can also be used for groundwater monitoring and give information about potential risks. Turbidity peaks in groundwater may originate from soil material, but also from the rapid ingress of surface water, run-off or surface percolate. Temperature measurements at different depths in boreholes can give information about the characteristics of the major inflows and rapid changes in temperature call for further investigation, such as examination for *E. coli, C. perfringens,* faecal streptococci and bacteriophages. This is especially important in poorly confined karst and fractured bedrock aquifers, where it is very difficult to predict flow paths. The major virtue of turbidity and temperature measurements is that they can be used on-line and in individual wells. If these indicate a risk event, they can be used to trigger both further investigations and control measures (shutdown of well, redirection of abstracted water, increased treatment).

### 4.5.2.2 *Pathogens as self-indicators*

Several groundwater surveys for pathogens have been conducted in the USA. Keswick and Gerba (1980) found enteric viruses in groundwater wells in concentrations of 1.2 plaque forming units (pfu)/l. In a nation-wide study in the USA, 150 wells from various States and aquifer types (70 consolidated, 34 bedrock, 46 unknown) were sampled and analysed for the presence of viruses and bacteriophages (Abbaszadegan *et al.,* 1999). Enteroviruses were found in 8.7% of the samples with cell culture and 27% of the samples using polymerase chain reaction (PCR). Hepatitis A was found in 8.0% and rotavirus was found in 12% of the samples (both using PCR methods). No correlation was found between enterovirus detected by cell culture and any of the other microbial indicators tested. These high prevalence rates indicate that virus contamination of groundwater is a greater public health concern than was formerly anticipated (Sobsey, 1999).

*Cryptosporidium* sp. and *Giardia* sp. have also been used to assess the contamination of groundwater systems. Hancock *et al.* (1997) found *Cryptosporidium* sp. and/or *Giardia* sp. in 12% of 463 groundwater samples from 199 sites in 23 American States. Infiltration galleries, adits and horizontal wells were most frequently contaminated with these parasites, but they were also found in springs and vertical wells. The mean concentration of *Giardia* sp.

cysts in positive wells was 8.4/100 l (maximum 120/100 l). For *Cryptosporidium* sp., the mean concentration was 5.1/100 l with a maximum of 45/100 l.


**4.6      Summary and outlook**

This chapter has illustrated the different ways to:

- Localise and characterise the sources of faecal contamination in a catchment area; and

- Determine the (variability of the) microbial quality of source water

of both surface water and groundwater systems. The objective of these activities is to collect information to substantiate and support (cost-effective) approaches to management of the risk of waterborne disease.

Several developments in recent years are providing new tools to help in the localisation of contamination sources and especially in understanding the significance of the sources. One of these developments is the integration of hydrological modelling and microbiology to construct transport models that describe and predict the fate of pathogens in both surface water and groundwater catchments (Schijven and Hassanisadeh, 2000; Deere *et al.*, 2000). This area is still in the early stages of development, but may significantly improve our understanding of the fate and 'behaviour' of pathogens and microbial parameters in the aquatic environment. The use of these models also forces researchers to focus more on the processes that govern transport and fate, rather than descriptive research on occurrence. A related development is the application of geographical information systems to localise sources and to relate contamination to waterborne outbreaks (Rose *et al.*, 2000) or to drinking water contamination events.

Microbial and non-microbial parameters provide a wide range of possibilities for measuring water quality changes and for the detection of faecal pollution. When these parameters have been used to their limit, pathogen detection may provide interesting, but difficult to interpret, data (Allan *et al.*, 2000). Developments in molecular microbiology (see Chapters 7 and 8) have provided methods that make it possible to detect these pathogenic microorganisms in water even those, such as Norwalk-like viruses, that could not previously be detected and analysed. These methods can be more specific and allow the differentiation of pathogenic and non-pathogenic strains and also allow genetic fingerprinting which is useful for identifying sources of contamination.

Similarly, the developments in (computer) technology have improved the automation of on-line measurements of water flow, temperature and turbidity. This is particularly helpful for rapid response, as illustrated by the case study from Melbourne Water, where this type of monitoring is used to divert the flow of contaminated streams away from the reservoir during storm events.

These new tools are refining the information available from monitoring surveys and they are also refining and accelerating the ability to respond to adverse conditions in a catchment or source water. It should not be forgotten, however, that much of the information that is needed to design a catchment protection strategy can be derived from a sanitary survey and that risk events can be deduced from simple parameters such as rainfall, river flow and turbidity. This information can be refined by monitoring for a set of microbial indices of faecal pollution or specific pathogens to obtain data on their occurrence in source water for the design of cost-effective treatment systems.

# REFERENCES

Abbaszadegan, M., Stewart, P. and LeChevallier, M. (1999) Occurrence of viruses in groundwater and groundwater disinfection rule: a nationwide study. Proceedings of the Water Quality and Technology Conference, Tampa, USA.

Allen, M.J., Clancy, J.L. and Rice, E.W. (2000) Pathogen monitoring – old baggage from the last millennium. *Journal of the American Water Works Association* **92**(9), 64-76.

Atherholt, T.B., Le Chevallier, M.W., Norton, W.D. and Rosen, J.S. (1998) Effect of rainfall on *Giardia* and *Cryptosporidium*. *Journal of the American Water Works Association* **90**, 66-80.

Bales, R.C., Li, S., Maguire, K.M., Yahya, M.T., Gerba, C.P. and Harvey, R.W. (1995) Virus and bacteria transport in a sandy aquifer, Cape Cod, MA. *Ground Water* **33**, 653-661.

Bales, R.C., Li, S., Yeh, T.C.J., Lenczewski, M.E. and Gerba, C.P. (1997) Bacteriophage and microsphere transport in saturated porous media: Forced-gradient experiment at Borden, Ontario. *Water Resources Research* **33**, 639-648.

Barbeau, B., Payment, P., Coallier, J., Clément, B. and Prévost, M. (2001) Evaluating the risk of infection from the presence of *Giardia* and *Cryptosporidium* in drinking water. *Quantitative Microbiology* **2**, 37-54.

Bitton, G. and Farrah, S.R. (1980) Viral aspects of sludge application to land. *American Society of Microbiology News* **46**, 622-624.

Blanc, R. and Nasser, A. (1996) Effect of effluent quality and temperature on the persistence of viruses in soil. *Water Science and Technology* **33**, 237-242.

Block, J.C., Joret, J.C., Morlot, M., and Foliguet, J.M. (1978) Recherche des enterovirus dans les eaux superficielles par adsorption elution sur microfibre de verre. *Tech. Sci. Munic.* **73**, 181-185.

Bouchier, I. (1998) *Cryptosporidium* in water supplies. Third Report of the Group of Experts, Dept. of the Environment, Transport and the Regions, London.

Byappanahalli, M.N. and Fujioka, R.S. (1998) Evidence that tropical soil environment can support the growth of *Escherichia coli. Water Science and Technology* **38**(12), 171-174.

Casemore, D.P., Wright, S.E. and Coop, R.L. (1997) Cryptosporidiosis – Human and animal epidemiology. In: *Cryptosporidium and Cryptosporidiosis*. Fayer, R. (Ed.) CRC Press, Boca Raton, USA. pp. 65-92.

Colwell, R.R. and Grimes, D.J. (2000) *Viable but Nonculturable Microorganisms in the Environment*. ASM Press, Washington, DC.

Craun, G.F., Hubbs, S.A., Frost, F., Calderon, R.L. and Via, S.H. (1998) Waterborne outbreaks of cryptosporidiosis. *Journal of the American Water Works Association* **90**, 81-91.

Curriero, F.C., Patz, J.A., Rose, J.B. and Lele, S. (2001) The association between extreme precipitation and waterborne disease outbreaks in the United States, 1948-1994. *American Journal of Public Health* **91**(8), 1194-1199.

Damgaard-Larsen, S., Jensen, K.O., Lund, E. and Nissen, B. (1977) *Water Research* **11**, 503.

DeBorde, D.C., Woessner, W.W., Kiley, Q.T. and Ball, P.N. (1999) Rapid transport of viruses in a floodplain aquifer. *Water Research* **33**, 2229-2238.

DeBorde, D.C., Woessner, W.W., Lauerman, B. and Ball, P.N. (1998) Virus occurrence in a school septic system and unconfined aquifer. *Ground Water* **36**, 825-834.

Deere, D., Davison, A., Stevens, M., Hellier, K., Mullenger, J. and Helm, G. (2000) Teaching old dogmas new tricks – the Australian approach to microbiological risk mitigation. Presented at the first World Water Congress, IWA, Paris.

DeReignier, D.P., Cole, L., Schupp, D.G., Erlandsen, S.L. (1989) Viability of *Giardia* cysts suspended in lake, river and tap water. *Applied and Environmental Microbiology* **55**, 1223-1229.

DWI (1998) Assessment of water supply and associated matters in relation to the incidence of cryptosporidiosis in west Herts and north London in February and March 1997. Drinking Water Inspectorate, London.

Dykes, A.C., Juranek, D.D., Lorenz, R.A., Sinclair, S., Jakubowski, W. and Davies, R.B. (1980) Municipal waterborne giardiasis: an epidemiological investigation. *Annals of Internal Medicine* **92**, 165-170.

EC (1975) Directive concerning the quality required for surface water intended for the abstraction of drinking water in the Member States. 75/440/EEC. *Official Journal* **L194**, European Commission, Brussels, Belgium.

Erlandsen, S.L. (1994) Biotic transmission – is giardiasis a zoonosis?. In: *Giardia from Molecules to Disease*. Thompson, R.C.A., Reynoldson, J.A. and Lymbery, A.J. (Eds.) CAB Int., Wallingford, UK. pp. 83-97.

Fayer R., Speer C.A. and Dubey, J.P. (1997) The general biology of *Cryptosporidium*. In: *Cryptosporidium and Cryptosporidiosis*. Fayer R (Ed.) CRC Press, Boca Raton, USA. pp. 1-42.

Fayer, R., Speer, C.A. and Dubey, J.P. (1990) In: *Cryptosporidiosis of Man and Animals*. Dubey, J.P. Speer, C.A. and Fayer, R. (Eds.) CRC Press, Boca Raton, USA.

Feachem, R.G., Bradley, D.J., Garelick, H. and Mara, D.D. (1983) Sanitation and disease: health aspects of excreta and wastewater management. In: *World Bank Studies in Water Supply and Sanitation 3*. Wiley, Chichester, UK.

Fenlon, D.R. (1981) Seagulls (*Larus* spp.) as vectors of salmonellae: an investigation into the range of serotypes and numbers of salmonellae in gull faeces. *Journal of Hygiene (London)* **86**(2), 195-202.

Feuerpfeil, I. and Bischoff, K. (2001) Ergebnisse von bakteriologischen und parasitologischen Untersuchungen an Trinkwassertalsperren in Sachsen und Thuringen. Proc. Statusseminar Zur Bedeutung mikrobiologischer

Belastungen fur die Trinkwasserversorgung aus Talsperren-Ein Zwischenbilanz. 14-15 Sep. 2000, Hennef, Germany, ATT-Schriftenreihe Band 2, Siegburg, Germany, pp.73-106.

Fujioka, R.S. (2001) Monitoring coastal marine waters for spore-forming bacteria of faecal and soil origin to determine point from non-point source pollution. *Water Science and Technology* **44**(7), 181-188.

Fujioka, R.S. and Yoneyama, B.S. (2001) Assessing the vulnerability of groundwater sources to fecal contamination. *Journal of the American Water Works Association* **93**(8), 62-71.

Funderburg, S.W. and Sorber, C.A. (1985) Coliphages as indicators of enteric viruses in activated sludge. *Water Research* **19**, 547-555.

Geldreich, E.E., Nash, H.D., Spino, D.F. and Reasoner, D.J. (1980) Bacterial dynamics in a water supply reservoir: a case study. *Journal of the American Water Works Association* **72**, 31-40.

Geldreich, E. E. (1970) Applying bacteriological parameters to recreational water quality. *Journal of the American Water Works Association* **62**, 113-120.

Geldreich, E.E. (1990) Microbiological quality of source waters for water supply. In: *Drinking Water Microbiology*. McFeters, G.A. (Ed.) Springer Verlag, New York, USA.

Geldreich, E.E. (1996) The worldwide threat of waterborne pathogens. In: *Water Quality in Latin America. Balancing the Microbial and Chemical Risks in Drinking Water Disinfection*. Craun, G.F. (Ed.) ILSI Press, Washington DC, USA, pp. 19-43.

Gerba C.P. and Bitton, G. (1984) Microbial pollutants: their survival and transport pattern to groundwater. In: *Groundwater Pollution Microbiology*. Bitton, G. and Gerba, C.P. (Eds.) John Wiley and Sons, New York, pp. 65-88.

Gerba, C. P. (1984) Applied and theoretical aspects of virus adsorption to surfaces. *Advances in Applied Microbiology* **30**, 133-168.

Gerba, C. P. *et al.* (1978) Characterisation of sewage solid-associated viruses and behaviour in natural waters. *Water Resources Research* **12**, 805-812.

Gibson III, C.J., Stadterman, K.L., States, S. and Sykora, J. (1998) Combined sewer overflows: a source of *Cryptosporidium* and *Giardia*? *Water Science and Technology* **38**, 67-72.

Grabow, W.O.K., Holtzhausen, C.S. and DeVilliers, J.C. (1993) Research on bacteriophages as indicators of water quality. Water Research Commission, Pretoria, South Africa. Project Report 321, 147pp.

Graczyk, T.K., Fayer, R., Trout, J.M., Lewis, E.J., Farley, C.A., Sulaiman, I. and Lai., A.A. (1998) *Giardia* sp. and infectious *Cryptosporidium parvum* oocysts in the feces of migratory Canada Geese (*Branta canadensis*). *Applied and Environmental Microbiology* **64**, 2736-2738.

Graczyk, T.K., Cranfield, M.R., Fayer, R. and Anderson, M.S. (1996) Viability and infectivity of *Cryptosporidium parvum* oocysts are retained upon intestinal passage through a refractory avian host. *Applied and Environmental Microbiology* **62**, 3234-3237.

Grimes, D.J. (1980) Bacteriological water quality effects of hydraulically dredging contaminated Upper Mississippi River bottom sediment. *Applied and Environmental Microbiology* **39**, 782-789.

Grimes, D.J. (1982) Bacteriological water quality effects of clamshell dredging. *Journal of Freshwater Ecology* **1**, 407-419.

Hagedorn, C. (1984) Microbiological aspects of groundwater pollution due to septic tanks. In: *Groundwater Pollution Microbiology*. Bitton, G. and Gerba, C.P. (Eds.) John Wiley and Sons, New York, pp. 65-88.

Hancock, C.M., Rose, J.B. and Callahan, M. (1997) The prevalence of *Cryptosporidium* and *Giardia* in US groundwaters. Proceedings of the International Symposium on Waterborne Cryptosporidium, March 1997, Newport Beach, USA, AWWA pp. 147-152.

Hansen, J.S. and Ongerth, J.E. (1991) Effects of time and watershed characteristics on the concentration of *Cryptosporidium* oocysts in river water. *Applied and Environmental Microbiology* **57**, 2790-2795.

Havelaar. A.H., Furuse, K. and Hogeboom, W.M. (1986) Bacteriophages and indicator bacteria in human and animal faeces. *Journal of Applied Bacteriology* **60**, 255-262.

Havelaar, A.H., van Olphen, M. and Drost, Y.C. (1993) F-specific RNA bacteriophages are adequate model organisms for enteric viruses in fresh water. *Applied and Environmental Microbiology* **59**, 2956-2962.

Havelaar, A.H. (1996) The place of microbiological monitoring in the production of safe drinking water. In: *Safety of Water Disinfection: Balancing Chemical and Microbial Risks.* Craun, G.F. (Ed.) ILSI Press, Washington DC, USA, pp.127-144.

Keswick, B.H. (1984) Sources of groundwater pollution. In: *Groundwater Pollution Microbiology.* Bitton, G. and Gerba, C.P. (Eds.) John Wiley & Sons, New York, pp. 39-64.

Keswick, B.H. and Gerba, C.P (1980) Viruses in groundwater. *Environmental Science and Technology* **14**,1290-1297.

Koenraad, P.M.F.J., Hazeleger, W.C., van der Laan, T., Beumer, R.R. and Rombouts, F.M. (1994) Prevalence of *Campylobacter* in Dutch sewage purification plants. *Food Microbiology* **11**, 65-73.

LaBelle, R.L. and Gerba, C.P.(1980) Relationships between environmental factors, bacterial indicators, and the occurrence of enteric viruses in estuarine sediments. *Applied and Environmental Microbiology* **39**, 588-596.

LaLiberte, P. and Grimes, D.J. (1982) Survival of *Escherichia coli* in lake bottom sediment. *Applied and Environmental Microbiology* **43**, 623-628.

Lassen, J. (1972) *Yersinia enterocolitica* in drinking water. *Scandinavian Journal of Infectious Disease* **4**,125-127.

LeChevallier, M.W. and Norton, W.D. (1995) *Giardia* and *Cryptosporidium* in raw and finished water. *Journal of the American Water Works Association,* **87**, 54-68.

Lucena, F., Finance, C., Jofre, J., Sancho, J. and Schwartzbrod, L. (1982) Viral pollution determination of superficial waters (river water and seawater) from the urban area of Barcelona (Spain). *Water Research* **16**, 173-177.

MacKenzie, W.R., Hoxie, N.J., Proctor, M.E., Gradus, M.S., Blair, K.A., Peterson, D.E., Kazmierczak, J.J., Addiss, D.G., Fox, K.R., Rose, J.B. and David, J.P. (1994) A massive outbreak in Milwaukee of

*Cryptosporidium* infection transmitted through the public water supply. *New England Journal of Medicine* **331**(3), 161-167.

Mara, D. and Cairncross, S. (1989) *Guidelines for the safe use of wastewater and excreta in agriculture and aquaculture.* World Health Organization, Geneva.

Mathess, G., Pekdeger, A. and Schroeter, J. (1988) Persistence and transport of bacteria and viruses in groundwater - a conceptual evaluation. *Journal of Contamination and Hydrology* **2**, 171-188.

McLellan, P. (1998) Sydney Water Inquiry. Fifth and final report. New South Wales, Premier's Dept., Sydney, Australia.

McNabb, J.F. and Mallard, G.E. (1984) Microbiological sampling in the assessment of groundwater pollution. In: *Groundwater Pollution Microbiology.* Bitton, G. and Gerba, C.P. (Eds.) John Wiley & Sons, New York, pp. 235-260.

Medema, GJ. (1999) Thesis: *Cryptosporidium and Giardia: New Challenges to the Water Industry.* University of Utrecht, Utrecht, The Netherlands.

Medema, G.J., Bahar, M. and Schets, F.M. (1997) Survival of *Cryptosporidium parvum, Escherichia coli,* faecal enterococci and *Clostridium perfringens* in river water: Influence of temperature and autochthonous microorganisms. *Water Science and Technology* **35**(11-12), 249-252.

Medema, G., Ketelaars, H. and Hoogenboezem, W. (2000a) *Cryptosporidium* and *Giardia:* vóórkomen in rioolwater, mest en oppervlaktewater met zwem- en drinkwaterfunctie. RIWA/RIVM/RIZA/Kiwa rapport. ISBN 9036953324. Report in Dutch.

Medema, G.J., Juhasz-Holterman and Luijten, J. (2000b) Removal of microorganisms by bank filtration in a gravel sand soil. Proceedings of the International Riverbank Filtration Conference, November 2000, Dusseldorf, Germany.

Medema, G.J., Ketelaars, H.A.M., Hoogenboezem, W. and Schijven, J. (2000c) *Cryptosporidium* en *Giardia*: het probleem, de oorzaken en de beheersing. $H_2O$ **33**(23), 31-34.

Moore, G.T., Cross, W.M., McGuire, D., Molahan, C.S., Gleason, N.N., Nealy, G.R. and Newton, L.H. (1969) Epidemic giardiasis in a ski resort. *New England Journal of Medicine* **281**, 402-407.

Morris, B.L. and Foster, S.S.D. (2000) *Cryptosporidium* contamination hazard assessment and risk management for British groundwater sources. *Water Science and Technology* **41**(7), 67-78.

Moulton-Hancock, C., Rose, J.B., Vasconcelos, G.J., Harris, S.I., Klonicki, P.T. and Sturbaum, G.D. (2000) *Giardia* and *Cryptosporidium* occurrence in groundwater. *Journal of the American Water Works Association* **92**, 117-123.

Nasser, A.M., Tchorch, Y. and Fattal, B. (1992) Comparative survival of *E. coli*, F⁺bacteriophages, HAV and poliovirus 1 in wastewater and groundwater. *Water Science and Technology* **27**, 401-407.

Nestor, L., Lazar, L., Sourea, D. and Ionescu, N. (1981) Investigations on viral pollution in the Romanian section of the Danube river during 1972-1977 period. *Zbl. Bakt. Hyg. I. Abt. Orig. B,* **173**, 517-527.

Olson, B. (1996) Pathogen occurrence in source waters: factors affecting survival and growth. In: *Safety of Water Disinfection: Balancing Chemical and Microbial Risks*. Craun, G.F. (Ed.) ILSI Press, Washington DC, USA, pp.83-98.

Payment, P. (1981) Isolation of viruses from drinking water of the Pont-Viau water treatment plant. *Canadian Journal of Microbiology* **29**, 111-119.

Payment, P., Fortin, S. and Trudel, M. (1986) Elimination of human enteric viruses during conventional waste water treatment by activated sludge. *Canadian Journal of Microbiology* **32**, 922-925.

Payment, P., Berte, A., Prevost, M., Ménard, B. and Barbeau, B. (2000) Occurrence of pathogenic microorganisms in the Saint-Lawrence river (Canada) and comparison of health risks for populations using it as their source of drinking water. *Canadian Journal of Microbiology* **46**, 565-576.

Pieper, A.P., Ryan, J.N., Harvey, R.W., Amy, G.L., Illangasekare, T.H. and Metge, D.W. (1997) Transport and recovery of bacteriophage PRD1 in a sand and gravel aquifer: Effect of sewage-derived organic matter. *Environmental Science and Technology* **31**, 1163-1170.

Poulton, M., Colbourn. J. and Dennis, P.J. (1991) Thames Water's experience with *Cryptosporidium*. *Water Science and Technology* **24**, 21-26.

Rivera, S.C., Hazen, T.C. and Toranzos, G.A. (1988) Isolation of fecal coliforms from pristine sites in a tropical rain forest. *Applied and Environmental Microbiology* **54**(2), 513-517.

Robertson, L.J., Campbell, A.T. and Smith, H.V. (1992) Survival of *Cryptosporidium parvum* oocysts under various environmental pressures. *Applied and Environmental Microbiology* **58**, 3494-3500.

Rolland, D., Hartemann, P., Joret, J.C., Hassen, A. and Foliguet, J.M. (1983) Evaluation of the load of enteroviruses in a biological waste water treatment plant. *Water Science and Technology* **15**, 115-121.

Rose, J.B., Daeschner, S., Easterling, D.R., Curriero, F.C., Lele, S. and Patz, J.A. (2000) Climate and waterborne outbreaks in the US: a preliminary descriptive analysis. *Journal of the American Water Works Association* **92**(9), 77-87.

Schijven, J.F. and Hassanisadeh, S.M. (2000) Removal of viruses by soil passage: overview of modelling, processes and parameters. *Critical Reviews in Environmental Science and Technology* **31**, 49-125.

Schijven, J.F. and Rijs, G. (2000) Rioolwater en mest als emissiebronnen voor *Cryptosporidium* en *Giardia*. *H₂O,* **33**(23), 25-27.

Schoenen, D. (2001) Analyse und Bewertung trinkwasserbedingter erkrankungen durch Parasiten. Proc. Statusseminar Zur Bedeutung mikrobiologischer Belastungen fur die Trinkwasserversorgung aus Talsperren-Ein Zwischenbilanz. 14-15 Sep. 2000, Hennef, Germany, ATT-Schriftenreihe Band 2, Siegburg, Germany, pp. 49-69.

Sobsey, M.D., Shields, P.A., Hauchman, F.H., Hazard, R.L. and Caton, L.W. (1986) Survival and transport of hepatitis A virus in soils, groundwater and wastewater. *Water Science and Technology* **18**, 97-106.

Sobsey, M. (1999) Monitoring for faecal viruses in ground water Proceedings of the Water Quality and Technology Conference, November 1999, Tampa, USA, AWWA.

Stewart, M.H., Ferguson, D.M., DeLeon, R. and Taylor, W.D. (1997) Monitoring program to determine pathogen occurrence in relationship to storm events and watershed conditions. Proceedings of the Water Quality and Technology Conference, November 1999, Tampa, USA, AWWA.

Tartera, C., Jofre, J. and Lucena, F. (1988) Relationship between numbers of enteroviruses and bacteriophages infecting *Bacteriodes fragilis* in different environmental samples. *Environmental Technology Letters* **9**, 407-410.

Tartera, C., Lucena, F. and Jofre, J. (1989) Human origin of *Bacteriodes fragilis* bacteriophages present in the environment. *Applied and Environmental Microbiology* **55**, 2696-2701.

Traverso, H.P. (1996) Water and health in Latin America and the Caribbean: infectious waterborne diseases. In: *Water Quality in Latin America. Balancing the Microbial and Chemical Risks in Drinking Water Disinfection.* Craun, G.F. (Ed.) ILSI Press, Washington DC, USA, pp. 45-54.

van Donsel, D.J. and Geldreich, E.E. (1971) Relationsips of salmonella to fecal coliforms in bottom sediments. *Water Research* **5**, 1079-1087.

van Olphen, M., Kapsenberg, J.G., van der Baan, E. and Kroon, W.A. (1984) Removal of enteric viruses from surface water at eight waterworks in the Netherlands. *Applied and Environmental Microbiology* **47**, 927-932.

Vaughn, J.M., Landry, E.F., Baranosky, L.J. Beckwith, C.A., Dahl, M.C. and Delihas, N.C. (1978) *Applied and Environmental Microbiology,* **36**, 47.

Wellings, F.M., Lewis, A.L., Mountain, C.W. and Stark, L.M. (1974) Virus survival following wastewater spray irrigation of sandy soils. In: *Virus Survival in Water and Wastewater Systems.* Malina, J.F. and Sagik, B.P. (Eds.) University of Texas, Austin, United States, pp. 253-260.

Whitlock, E.A. (1954) The waterborne diseases of microbial origin. Proceedings of the Annual Conference of the National Association of Bath Superintendents. Anderson Ltd, Stepney Green, UK, p.3-19.

WHO (1989) *Guidelines for the Safe Use of Wastewater and Excreta in Agriculture and Aquaculture.* World Health Organization, Geneva.

WHO (1997) *Guidelines for drinking water quality. Second Edition. Volume 3. Surveillance and control of community supplies Second Edition.* World Health Organization, Geneva.

WRc (1991) Surrogate viral indicators. Water Research Centre, Marlow. Report no. FR/D 0005, Foundation for Water Research, UK.

Woo, P.K. (1984) Evidence for animal reservoirs and transmission of *Giardia* infection between species. In: *Giardia and Giardiasis*. Erlandsen, S.L. and Meyer, E.A. (Eds.) Plenum Press, New York, USA, pp. 341.

*Chapter 5*

## TREATMENT EFFICIENCY

*G. Stanfield, M. Lechevallier and M. Snozzi*


## 5.1    Introduction

The primary purpose of water treatment is to provide drinking water to consumers that is free of waterborne pathogens. Because no single treatment process can be expected to remove all of the different types of pathogens that can be found in water (under all conditions), multiple barriers are desirable. Multiple barriers will also ensure additional safety in the case that a single treatment step is not working optimally. The number of treatment processes (technical barriers) required is influenced by the quality of the source water (see Chapter 4). Groundwaters that are protected from surface influence are usually of relatively good quality and so traditionally few, if any, treatment processes are required. Lowland surface water sources are usually of much poorer quality and more treatment processes are needed to provide an acceptable level of safety.

A number of treatment processes are also designed to modify the chemical and physical properties of the water (rather than pathogen elimination). State of the art treatment includes techniques to reduce AOC and reducing matter, so that on the one hand the regrowth of the pathogens in the distribution system is low and on the other hand the disinfection is more effective. This chapter, however, does not detail such processes but concentrates solely on the reduction of faecal-oral infection risk.

A wide spectrum of pathogenic agents can be found in water and monitoring for their presence on a routine basis is impracticable. Traditionally (as outlined in Chapter 1) microbial safety of drinking water has been confirmed by monitoring for the absence of microorganisms of faecal origin. Bacteria such as *E. coli*, faecal streptococci and *Clostridia* have been used for this purpose, because they are consistently present in high numbers in the faeces of warm-blooded animals and are relatively easy to detect in water. These bacteria and groups of bacteria are microbial indices of faecal pollution and form the basis of guidelines and national standards.

It has been recognised that the microbial indicator parameters do not necessarily behave in the same way as certain pathogens in water treatment processes. The ability of treatment processes to remove specific pathogens has been directly measured, with such studies typically conducted at bench or pilot scale some of them using water spiked with pathogens (Sommer and Cabaj, 1993; Jacangelo *et al.*, 1995; Bellamy *et al.*, 1985; Hunt and Mariñas, 1999). The potential removal determined in such pilot studies will, however, not necessarily be achieved in full-scale treatment. Therefore, there is a need for alternative parameters that correlate more closely with the behaviour of specific pathogens both to assess the disinfection potential of full-scale treatment and to measure process performance during treatment.

Safe drinking water is the result of careful evaluation of source water quality and variation (as outlined in Chapter 4) and adequate, reliable treatment processes combined with performance monitoring to assure that treatment is within operating parameters. The focus for the control of process operation should be put on simple measurements, which can be done on-line. If the input to the system and its normal performance is known, the on-line measurement will be a perfect indication of disturbances and changes in the water quality. This shifts the emphasis of quality control of drinking water from end product testing (*i.e.* testing for failure) to the testing and control of treatment processes (*i.e.* preventing failure). Current treatment processes and appropriate indicators of performance are discussed below.

A verification of the quality at the end of the treatment chain is necessary. For this purpose non microbial parameters like flow, colour and disinfectant residual (where appropriate) are suitable (see also Chapter 2, Table 2.4). Microbial parameters for the validation of the treatment process include *E. coli*, total and thermotolerant coliforms, heterotrophic bacteria and aerobic spore-forming bacteria. However, it should be stressed that this verification should not be mistaken as a determination of the safety of the drinking water.

## 5.2 Microbial treatment efficiency

A review of the available data on treatment efficiencies has been published by LeChevallier and Au (2002). Disinfection can be achieved in two ways:

- The physical removal of the pathogens.

- The inactivation (death) of the pathogen.

Apart from careful characterisation of the disinfection potential of a given treatment process (which in many cases involves experimental determinations) it is also important to identify simple measurements that give information rapidly on whether the treatment process is working properly. For the latter, physical and chemical measurements (preferably on-line measurements) are often better than microbial determinations.

A review of potential inactivation rates for different disinfection treatments has been published by Sobsey (1989). More recently, the United States Environmental Protection Agency (USEPA, 1999) has compiled data on achievable disinfection efficiencies for various processes and combinations of treatment steps. Although the actual inactivation will be influenced by many factors (including the ability of many of the microbial parameters to remain viable while becoming non-culturable), the following subsections list typical ranges reported for each treatment process. Although retention of water in reservoirs and impoundments can bring about significant improvements in quality as a result of inactivation, sedimentation and predation this process is not discussed here. For more details the reader is referred to the review by LeChevallier and Au (2002). For more precise evaluation of the reduction of individual pathogens by a treatment process, specific experimental determination is necessary.

### 5.2.1  Coagulation and sedimentation

The most common coagulants in use throughout the world are aluminium sulphate, ferric sulphate, ferric chloride and poly-aluminium chloride. These coagulants are mixed into the water where they produce hydroxide precipitates that are 'fluffy' and enmesh particles and microbes along with some of the dissolved organic carbon. In some circumstances, flocs generated by aluminium and ferric salts can be strengthened by the addition of coagulant aids such as long chain organic polymers. The flocs formed by this process must be removed. This can be achieved by sedimentation or, if the flocs are very light,

161

fine air bubbles may be used to carry them to the surface (air flotation) where they are skimmed off. They can also be removed by direct filtration.

Various forms of coagulation and sedimentation are used in water treatment and there are differences in general practices between countries, which makes the comparison of data difficult. However, published data indicate that this process may remove between 40% and 99% of bacteria, which translates into 0.2 and 2 logs of removal. Removal of viruses is rather poor, below 1 log, whereas for parasites such as *Cryptosporidium* removal of up to 2 logs has been reported.

The retention of the formed flocs is very important because of the accumulation of pathogens, since even single flocs may contain sufficient numbers of pathogens to be of hygienic importance (Gale *et al.*, 1997). Continuous measurements of turbidity or particle counts are useful for monitoring the efficiency of this process.

### 5.2.2    *Filtration*

Various filtration processes are used in drinking water treatment. Used with proper design and operation, filtration can act as a consistent and effective barrier against microbial pathogens. Filtration processes that are used in potable water treatment and the pore size of filter medium are shown in Figure 5.1, along with the sizes of selected microbial particles. This provides an insight into the removal mechanisms and likely efficiencies of the different filtration processes.

Filtration is a physical removal of organisms together with other particulate matter. On-line measurements of turbidity or particle counts, as well as determination of particle size distribution are excellent control parameters for this process. If parallel filtration units are operated, it is essential that each unit is measured separately in order to ensure the recognition of poor performance in an individual filter unit.

**Figure 5.1. Filter medium pore sizes and the size of microbial particles
(with selected microorganisms marked with numbers)**

(Adapted from LeChevallier and Au, 2002)

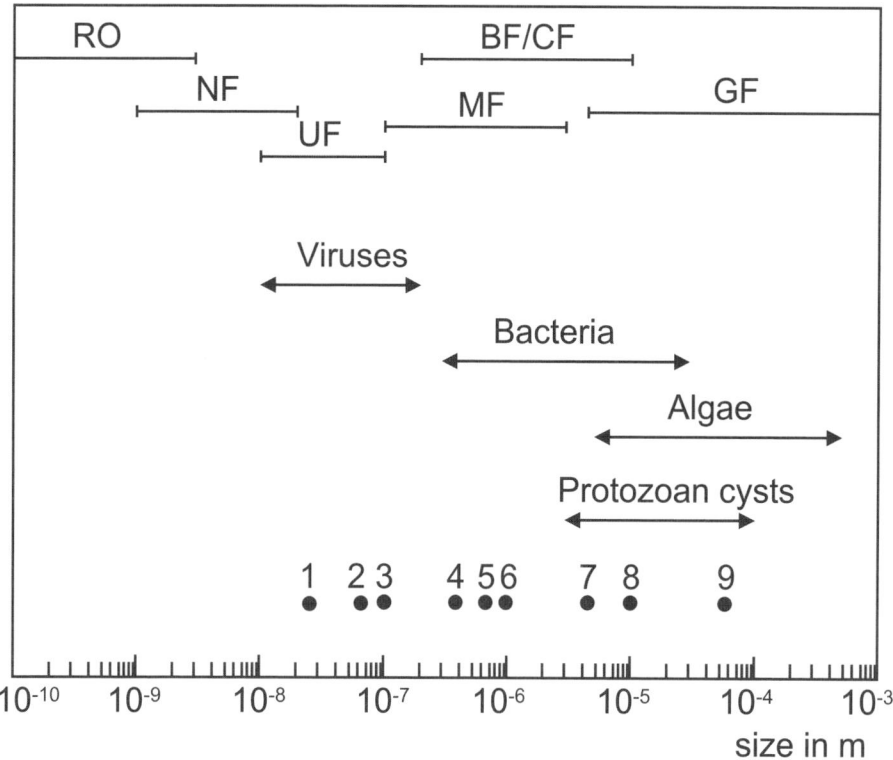

Key:
RO: reverse osmosis. NF: nanofiltration. UF: ultrafiltration. MF: microfiltration. BF/CF: bag and cartridge filters. GF: granular filtration including slow sand filtration (slow sand filters have lower pore sizes than rapid-rate filters)
1. MS2 bacteriophage. 2. Rotavirus. 3. PRDI bacteriophage. 4. *Mycobacterium avium* complex (represents smallest size). 5. *Yersinia* spp. 6. Coliform bacteria. 7. *Cryptosporidium* oocysts. 8. *Giardia* cysts. 9. *Balanthidium coli* cysts.

## 5.2.2.1   *Rapid filtration*

Rapid filters are deep beds (0.6-1.0 metres) of sand, anthracite and sand or granular activated carbon. The particle size of the medium is usually about 1 mm. They are operated at flow velocities of about 5-15 metres per hour. Rapid filters retain most of the flocs and other particles that escape chemical coagulation and sedimentation. The size of particles that can be removed in deep-bed filtration can be much smaller than the pore size of the filter (Hall,

1998). This is due to electrostatic adhesion causing adsorption of particles that are in close proximity to the filter medium. Rapid filters are stopped and backwashed according to a time cycle (usually 24 hours), when flow becomes excessively restricted due to clogging or when the turbidity or particle counts of the filtrate become unacceptably high.

The microbial removal efficiency of rapid filters can be influenced by a number of factors. Correct operation and maintenance of rapid filters is essential otherwise performance may be lost. In poorly maintained filters, cracks have been observed particularly near the walls, which allow unfiltered material to pass through, decreasing the bacteriological quality of the filtrate. Changes in the flow rate can dislodge deposits containing microorganisms causing them to pass into the filtrate. When a filter is put back into service after backwashing, the initial filtrate is of poor quality in terms of turbidity and bacterial numbers. This is due to displacement of residual backwash water, and the lower efficiency of the clean filter media, compared with a partly used (ripened) filter (Amirtharajah and Wetstein, 1980). For this reason the initial filtrate may be run to waste or returned to the start of the treatment processes for a period of up to 30 minutes. Alternatively a 'slow start' procedure may be used in which the flow rate through the filter is restricted until the filtrate becomes of acceptable quality. Additionally, backwash water should not be recycled within the treatment plant.

Published data indicate that coagulation combined with rapid filtration may remove between 2 and 3 logs of bacteria, while reported removal of viruses range from 1 to 3 logs and for parasites such as *Cryptosporidium* 2 to 3 logs. Continuous measurements of turbidity and/or particle counts are important for monitoring.

### 5.2.2.2   Slow sand filtration

Slow sand filtration is a biological treatment process, which has to be used without coagulation pre-treatment. Other pre-treatment, particularly rapid filtration, may be used to remove high particle loads. Typically, a slow sand filter has a depth of about 0.7 metres and is operated at flow rates of 0.1 to 0.3 metres/hour compared to 5-15 metres/hour in rapid filters. The sand is mixed in size ranging from 0.15 to 0.35 mm. The pores are still quite large at about 60 μm. Although there is some filtration in depth, as in rapid sand filtration, the vital process is the formation of a biologically active layer (the Schmutzdecke) in the top 20 mm. Optimum treatment performance is dependent on a well-established Schmutzdecke. This provides an effective surface filtration of very small particles, including bacteria, parasites and viruses.

Any particles that pass through the Schmutzdecke may be retained in the remaining depth of the sand by the same mechanisms as exist in rapid filtration. The growth of the Schmutzdecke and its retention of particles cause a loss of permeability in the top layer of sand so that after some weeks of operation, flow rates decline. When this occurs, the filter is taken out of service and the top 20-30 mm removed by skimming. Slow sand filters are known for their high efficiency in removing bacteria and parasites, but small channels can occur in the filter if not properly operated and maintained which influence performance. In well-maintained systems with slow sand filtration it is possible to achieve a performance similar to a combination of coagulation and filtration. Continuous measurements of turbidity and/or particle counts are important for monitoring.

### 5.2.2.3   Activated carbon filtration

Activated carbon filters are predominantly used to remove organic compounds. However, they may also affect counts of microbial organisms including reduction of viruses and parasites. Due to growth in the filters, increased heterotrophic plate counts and total coliform counts can sometimes be observed.

### 5.2.2.4   Membrane filtration

In membrane filtration water is passed through a thin film, which retains contaminants according to their size. Membrane filtration has been playing an increasing role in drinking water treatment, including pathogen removal. The most commonly used membrane processes in drinking water treatment for microbial removal are microfiltration (MF) and ultrafiltration (UF) (see Figure 5.1). Detailed description of the fundamentals, design and operation of these processes are available in the literature (AWWARF, 1996; Taylor and Wiesner, 1999). Other membrane processes such as reverse osmosis (RO) and nanofiltration (NF), which are used primarily for other purposes, also remove pathogens.

Membrane filtration removes microbial pathogens primarily by size exclusion; microbes with sizes greater than the membrane pore size are removed. Chemical coagulation prior to the membrane is not a requirement for microbe removal. However, some degree of pre-treatment must be employed to reduce membrane fouling. Fouling arises from accumulation of chemicals, particles and the growth of organisms on membrane surfaces, resulting in reduced membrane productivity. Once fouling accumulates to such a level that the productivity of the system is unacceptable, the membranes must be chemically cleaned to restore productivity. Advanced pre-treatment systems such as conventional coagulation-sedimentation-filtration or other membrane processes may also be considered, depending on the quality of the source water.

Published data indicate that membrane filtration may remove up to 6 logs of bacteria, viruses or parasites. Process performance is generally monitored by measurement of physical parameters such as pressure drops across the membrane.

### 5.2.3   Chemical inactivation

Chemical disinfection to inactivate pathogens is an important treatment barrier. Chemicals used include chlorine, chloramine, chlorine dioxide and ozone. Treatment effectiveness is a function of dose, contact time, temperature and sometimes pH. Chemical disinfection can be placed at different positions in the treatment train and more than one disinfectant can be used, however it is important to note that organisms entrapped in particles may be shielded from the action of the chemicals. Primary disinfection is the process by which microorganisms are inactivated during the treatment process, while a secondary disinfectant can be added prior to distribution to maintain the water quality within the distribution system. Secondary disinfection provides a final barrier against bacterial contamination and regrowth within the distribution system. The practice of residual disinfection is, however, controversial (IWSA, 1998). It has been suggested that if biological stability is achieved and the system is well maintained then the disinfectant is unnecessary and may mask ingress into the distribution system by killing the bacterial indicators (but not the more robust pathogen microorganisms).

The concept of disinfectant concentration and contact time is integral to the understanding of disinfection kinetics and the practical application of the CT concept (which is defined as the product of the residual disinfectant concentration [C in mg/l] and the contact time [T in minutes], that residual disinfectant is in contact with water – USEPA, 1999) is important. Allowance must be made for the decline in concentration over time and in measuring time it is important to take account of the hydraulic behaviour of the treatment plant (in particular any short-circuiting). Temperature, over the range appropriate for drinking water, affects the rate of disinfection reactions according to the Arrhenius Law, although some deviations have been noted for certain disinfectants at low temperatures. The pH of the disinfectant solution also affects reaction kinetics. Table 5.1 outlines CT values for inactivation of viruses.

**Table 5.1.  CT values for virus inactivation**

(USEPA, 1999)

| Disinfectant | Units | Inactivation | | |
| --- | --- | --- | --- | --- |
| | | 2-log | 3-log | 4-log |
| Chlorine[1] | mg min/l | 3 | 4 | 6 |
| Chloramine[2] | mg min/l | 643 | 1 067 | 1 491 |
| Chlorine dioxide[3] | mg min/l | 4.2 | 12.8 | 25.1 |
| Ozone | mg min/l | 0.5 | 0.8 | 1.0 |
| UV | mW s/cm$^2$ | 21 | 36 | not available |

1. Values based on a temperature of 10 °C, pH range 6 to 9, and a free chlorine residual of 0.2 to 0.5 mg/l.
2. Values based on a temperature of 10 °C and a pH of 8.
3. Values based on a temperature of 10 °C and a pH range of 6 to 9.

### 5.2.3.1  *Chlorination*

Chlorination can take a number of forms including the use of chlorine, chloramines and chlorine dioxide. Each chemical has different disinfecting properties. Monochloramine (formed by the combination of chlorine with nitrogenous compounds) has a lower disinfection activity than chlorine but is more stable. Chlorine dioxide may be chosen because of its greater effectiveness against parasites.

Nearly 100 years of drinking water chlorination has demonstrated its effectiveness in the inactivation of microbial pathogens and the benefits of chlorination out-weigh any disadvantages, such as production of trihalomethanes. Enteric viruses are generally more resistant to chlorine than enteric bacteria, and viruses associated with cellular debris or organic particles may require high levels of disinfection due to the protective nature of the particle surface. Chlorination is considered to be highly effective for virus inactivation if the water has a turbidity of ≤ 1.0 nephelometric turbidity units (NTU), a free chlorine residual of 1.0 or greater for at least 30 minutes, and a pH of < 8.0. Protozoan cysts such as those of *Cryptosporidium* and *Giardia lamblia*, however, are highly resistant to chlorine disinfection (USEPA, 1989). Other factors that influence microbial sensitivity to chlorine include surface attachment, encapsulation, aggregation and low-nutrient growth.

Chlorine is a strong disinfectant that is effective at inactivating bacteria and viruses and, under certain circumstances, Giardia. CT values for 2 log inactivation of vegetative bacteria may vary between 0.02 and 200 mg min/l

(Grohmann, A; 2002) This wide range depends on a number of factors particularly the presence of reducing matter. One purpose of water treatment, therefore, is to eliminate such matter from water prior to chlorination. Residual levels of reducing matter can be determined by electrochemical methods such as oxidation-reduction-potential (ORP) measurements. The use of high dosage of chlorine is, therefore, by itself not a guarantee of safe drinking water as the presence of reducing matter may result in high concentrations of disinfection by-products (DBP), such as trihalomethanes (THM), which are toxic.

No significant reduction of Cryptosporidium is achieved with conventional CT values. Since pH, temperature and chemical composition will influence the disinfection potential they need to be monitored together with the CT measurements.

Because of the weak disinfecting power of monochloramine, it is not recommended as a primary disinfectant and it is ineffective in the inactivation of *Cryptosporidium*. Most systems using monochloramine apply a short period of free chlorine prior to ammonia addition or use an alternative (*e.g.* ozone, chlorine dioxide) primary disinfectant. Chloramines have CT values of more than 80 mg min/l for a 2 log inactivation of bacteria; values for the same inactivation of viruses are above 600 mg min/l and, therefore, they are only suitable for the inactivation of bacteria.

Chlorine dioxide is a strong oxidant as well as a powerful disinfectant and, therefore, can be used for the control of iron, manganese and taste and odour causing compounds as well as a primary disinfectant. It has also been used as a secondary disinfectant in many European countries. However, chlorine dioxide forms inorganic by-products (chlorite and chlorate ions) upon reaction with water constituents, and a water supplier may need to provide additional treatment depending on the level of these inorganic by-products and specific regulatory requirements. Chlorine dioxide is roughly comparable to free chlorine for inactivation of bacteria and viruses at neutral pH (), but it is more effective than free chlorine at an alkaline pH of 8.5 (Hoff and Geldreich, 1981). CT values for chlorine dioxide resulting in a 2 log inactivation of vegetative bacteria are less than 1 mg min/l. While values around 4 mg min/l have been reported for viruses and those for *Giardia* inactivation are around 15 mg min/l. Temperature and chemical composition need to be monitored together with the CT measurements (or calculations) and chlorine residual.

Chlorination usually takes place at a central treatment point but, particularly in developing countries, there is growing interest in applying it at household level. Sachets or tablets of a chlorine compound (sometimes together with a coagulant to remove turbidity) are sometimes used. Decentralised production of sodium hypochlorite is now possible from the electrolysis of a solution of common salt and this may provide a cost-effective source of chlorine solution. Combined coagulant-disinfectant tablets or powders or use of a solution of sodium hypochlorite are available for household water treatment (Sobsey, 2002).


## 5.2.3.2   Ozonation

Ozone has been used for more than a century for water treatment, mostly in Europe, but this usage is spreading to other areas. Despite this long use, the exact mechanism of how ozone inactivates microbes is not well understood, although it is known that ozone in aqueous solutions may react with microbes by direct reaction with the molecular ozone, or via reaction with the radical species formed on ozone decomposition.

Of the vegetative bacteria, *E. coli* is one of the most sensitive to ozone disinfection, while Gram-positive cocci (*Staphylococcus* and *Streptococcus*), the Gram-positive bacilli (*Bacillus*) and the mycobacteria are the most resistant. *Mycobacterium avium* can be effectively controlled by low doses of ozone, whereas the organism is highly resistant to free chlorine. It has been reported that heterotrophic plate count bacteria may be less susceptible to ozone inactivation than other indicator organisms. Viruses are generally more resistant to ozone than vegetative bacteria, although phages appear to be more sensitive than human viruses. Ozone is effective against *Giardia* and to a lesser extent *Cryptosporidium*. Because ozone does not produce a stable residual it is frequently followed by chlorination to produce a residual disinfectant for distribution. Due to the relatively fast decay of ozone even in pure water, hydraulics of the ozonation reactor are very important (see below).

Ozone will oxidise organic components present in the water, such as natural organic matter to produce smaller organic substances. Since these are usually more biodegradable, ozonation will increase bacterial growth after treatment. To prevent this, post ozonation removal of the oxidation products is necessary.

Ozone is a very powerful disinfectant for inactivation of vegetative bacteria. CT values below 0.5 mg min/l are reported for 2 log reduction of bacteria. CT values between 0.5 and 1 mg min/l are required for a 2 log inactivation of viruses. Inactivation of protozoa like *Giardia* is possible at temperatures above 15°C with CT values of 0.7 mg min/l for 2 log inactivation, while at 5°C the CT value increases to 1.3 mg min/l. For the same inactivation of *Cryptosporidium* the CT values required are about ten times higher. Content of organic carbon will also influence the disinfection efficiency. Therefore the measurement of CT values needs to include control of temperature and quality of water entering the ozonation reactor.

*Case study: Hydraulics of an ozonation reactor*

In Switzerland, food-related laws and regulations require health risks assessment and the evaluation of critical treatment steps in drinking water production. In the city of Zurich, a considerable fraction of drinking water is produced from lake water following a multistage procedure. During a health risk assessment the hydraulics of the ozonation reactor were evaluated by addition of a concentrated sodium chloride solution to the water inlet of the reactor for a period of two hours. Five sampling points along the water flow allowed the spread of the addition through the reactor to be followed (Kaiser *et al.*, 2000). Modelling of the experimental data showed that the reactor was best described by a series of four mixed reactors followed by a plug-flow reactor with considerable back flow. The model was confirmed by the comparison of modelled and measured ozone profiles and atrazine concentrations.

Modelled inactivation of microorganisms showed a remarkable difference between a single plug-flow model and the model derived from the experimental measurements. According to the model, the ozonation should reduce vegetative bacteria and viruses by more than 6 logs, spores of *Bacillus subtilis* will be inactivated by 1.5 logs, whereas the inactivation of *Cryptosporidium* is less than 1 log.

*5.2.3.3    UV disinfection*

UV action results from absorption by nucleic acids (DNA and RNA), leading to the dimerisation of pyrimidine bases, and all organisms are susceptible to UV light. Exposure to UV results in reduced viability of the treated cells. However, most bacteria have evolved different repair systems to cope efficiently with UV damage to their genetic material, for example, thymine dimers can be repaired both in the presence ('photoreactivation') or absence of light ('dark repair') (Jagger, 1967). Thus, UV doses in a certain range will only transiently reduce the ability of bacteria to form colonies without having a long-term effect on their survival (Mechsner *et al.*, 1991). Therefore, for the UV disinfection of drinking water it is essential to treat each volume part with a sufficient light dose to kill the bacteria. Usually a dose of 400 J/m$^2$ (40 mW s/cm$^2$) is accepted as being sufficient for efficient treatment.

Three types of light source are used for UV disinfection, namely:

- Low-pressure mercury lamp.

- Medium-pressure mercury lamp.

- Pulsed lasers.

The most popular so far is the low-pressure mercury lamp, which emits light at the wavelength of 254.7 nm, almost exclusively. Due to the rather low light intensity of such lamps, radiation times required for efficient disinfection are substantially higher than those for the second type, the medium pressure mercury lamp, which emits light of higher intensity and also of longer wavelength. It is sometimes claimed that the medium-pressure lamps have a better performance, because they may act in a dual way, damaging both DNA and proteins, some of which might be involved in the DNA-repair process. On the other hand, due to the much higher light intensity of medium-pressure lamps, the required contact time is much shorter with a concomitant risk of volume parts not being treated sufficiently. Recently, the use of pulsed UV lasers has been suggested. It is claimed that the same extent of cell inactivation can be achieved with this light source at less than one tenth of the dose of low-pressure mercury lamps. Rubin *et al.* (1982) showed a dependence of photoinactivation of yeast cells on the UV light intensity at the same dose. A similar dependence was observed for the photoprotection. Therefore, at high light intensity more dead cells were found at lower doses.

Another factor interfering with this type of disinfection is the UV transmission of the water. For treatment process evaluation the minimal UV dose for water with different UV transmission characteristics must be known. Biodosimetric determination of the UV dose under production conditions has been proposed as the best method for determining efficiency (Sommer and Cabaj, 1993). This procedure includes the addition of spores of *Bacillus subtilis* to the water before treatment; from the difference between the colony counts before and after treatment the UV dose in the reactor can be inferred from a dose-response curve determined in the lab. Similar dose-response curves can be determined for other organisms of interest (*e.g.* pathogens) and the reduction potential of the treatment system can be evaluated.

The transmission of the water should be monitored on-line with the help of an UV detector. The determination of colony forming units of coliform bacteria is not a satisfactory measure of UV inactivation because of the possibility of repair mechanisms coming into play (Mechsner *et al.*, 1991). If a microbial indicator parameter is required, the reduction of spores should be measured since they are easy to measure and at the same time quite resistant to UV light.

UV disinfection has been proven to be adequate for inactivating bacteria and viruses. UV doses of 400 J/m$^2$ will reduce vegetative bacteria by 4 to 8 logs. Virus inactivation is by 3 to 6 logs. Protozoa are more resistant to UV disinfection, but newer studies showed that in neonatal mouse infection studies with UV treated *Cryptosporidium oocysts* at a UV dose of 410 J/m$^2$ a 4 log reduction in infection occurred. Similar UV doses are required for a 4 log reduction in spores of *Bacillus subtilis*.

## *UV disinfection case study*

In Austria, Germany and Switzerland certification requirements have been established for the UV disinfection of drinking water, which typically require biodosimetric determination of the disinfection efficiency under production conditions (Snozzi *et al.*, 1999). Spores of *Bacillus subtilis* are used for this process since repair mechanisms are not important and can be neglected. The water entering the UV plant is inoculated with the spores and their concentration is determined before and after UV treatment. The UV dose can be calculated from the reduction of viable spores and a dose-response curve measured previously in the lab. Variation of the light intensity and the flow rate allows the definition of the range of flow rate with turbulent mixing within the reactor.

The result of the experimental determination of the disinfection efficiency can be represented in a graph (Figure 5.2) showing the maximal flow rate as a function of the UV transmissions of the water, which will ensure a minimal radiation dose of 400 J/m$^2$. If operation remains within these limits, the predetermined reduction of the number of viable pathogens can be ensured.

This experimental determination of the UV disinfection potential of a given reactor is very reproducible. Deviations between different determinations several months apart were found to be less than 2% (Snozzi, 2000).

Measurements of UV light intensity in the reactor serves as a control for process performance (it is important that the measuring point should be positioned such that changes in the UV transmission of the water will influence the reading of the light meter).

**Figure 5.2.  Measured UV light intensity as a function of UV transmission of the water**

(Adapted from Snozzi, 2000)

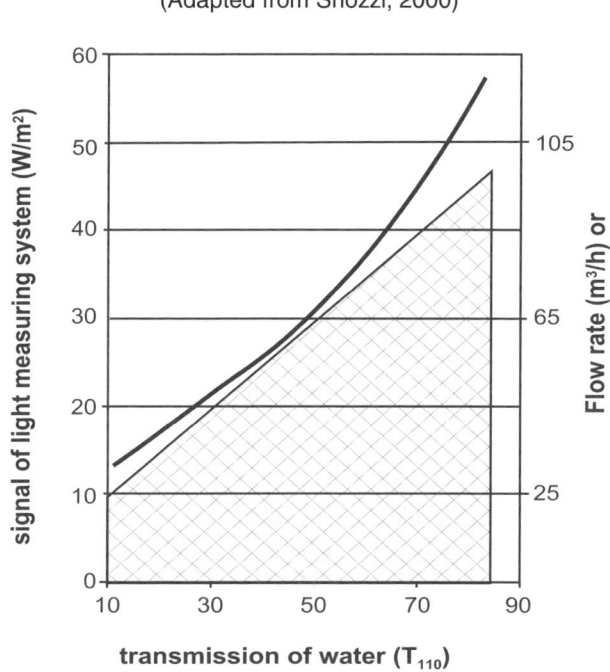

The dashed area represents combinations of UV transmissions and flow rates, which result in reliable disinfection. The solid curve represents the signal of the light measuring device as a function of the UV transmission of the water. UV transmission ($T_{100}$) is given as percent transmission using a 100 mm light pass.

### 5.2.3.4    Solar water disinfection

Solar panels can be used to generate electricity to power the UV lamps mentioned in the previous section but in low-income countries the sunlight alone can be used to kill or inactivate many, if not all, of the pathogens found in water. Solar water disinfection is a method of treating relatively small amounts of water at the point of use. There are three ways in which solar radiation can be used to eliminate pathogens. The first is through heating, the second through the effect of the natural UV radiation and the third through a mixture of both thermal and UV effects. None of these methods is yet widely used but laboratory experiments and field programmes show that some systems have good potential to produce potable water. Solar disinfection is included in the technologies reviewed by WHO for household water treatment and storage (Sobsey, 2002).

Thermal heating from the sun can be via solar cookers (which concentrate the rays of the sun with reflectors) or from simply exposing black-painted containers to the sun. In many systems temperatures can reliably reach over 55°C killing many pathogens. With the cookers and some of the other systems the temperature of the water can easily exceed 65°C, a pasteurisation temperature capable of inactivating nearly all enteric pathogens. Achievement of specific temperatures can be monitored using simple low-cost re-usable water pasteurisation indicators, based on the visible melting of wax in a clear plastic tube.

The use of heating and UV radiation to simultaneously disinfect water is used by a number of different solar treatment systems. The widest known is the SODIS system (Figure 5.3), which is suitable for low-income countries. The only equipment required is locally available bottles to contain the water (which needs to have a turbidity <30 NTU). This technique is now being field tested in various parts of the world and increasing amounts of data are becoming available on its effectiveness. Obviously for the UV to be effective the bottle material needs to be transparent to the useful wavelengths of the UV rays. The promoters of SODIS suggest the use of thin PET plastic bottles rather than PVC ones because the former material is more chemically stable. The half of the bottle furthest from the sun should be painted with black paint to improve the heat gain from the absorption of thermal radiation, and the bottle can be laid on a dark roof to further increase the potential temperature rise in the water. Shaking a partly filled bottle to aerate the water before filling it completely has been found to give a faster pathogen kill rate (Reed, 1997). The water requires several hours of exposure to strong sunlight to obtain the advantageous synergy between UV dosage and temperature rise (Wegelin et al., 1994, Sommer et al., 1997). In cloudy weather a much longer period (such as two days or more) is required because of the lower level of UV radiation and the reduced likelihood of the temperature of the water ever exceeding 50°C.

**Figure 5.3. Schematic representation of solar water disinfection and the influence of the water temperature on the UV-inactivation of bacterial cells**

(Printed with permission of M. Wegelin)

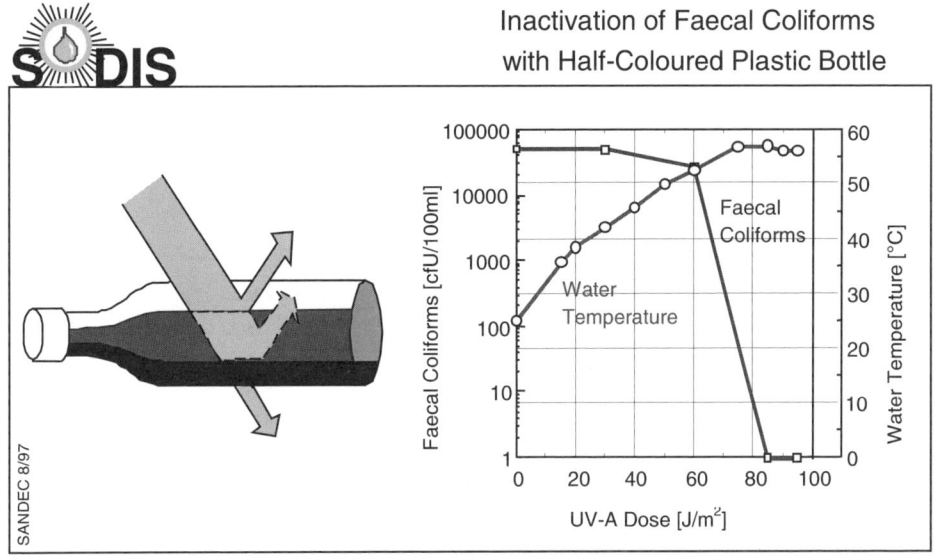

### 5.3    Summary

This chapter reviews the different treatment barriers available to ensure the production of safe drinking water. The choice of which barriers to implement depends on a number of considerations including the source water quality. Non-microbial indicator parameters that can be measured on-line are most useful for assessing process performance and such monitoring is important within the total system approach to risk management. Treatment steps with relevant pathogen removal or inactivation are described together with possible indicators for the measurement of process performance.

175

# REFERENCES

AWWARF (1996) *Water Treatment Membrane Processes*. American Water Works Association, Lyonnaise des Eaux, and Water Research Commission of South Africa. McGraw-Hill, Inc., New York.

Amirtharajah, A. and Wetstein, D.P. (1980) Initial degradation of effluent quality during filtration. *Journal of the American Water Works Association* **78**, 66-73.

Bellamy, W.D., Hendricks, D.W. and Logsdon, G.S. (1985) Slow sand filtration: influences of selected process variables. *Journal of the American Water Works Association* **77**, 62-66.

Gale, P., Van-Dijk, P.A.H. and Stanfield, G. (1997) Drinking water treatment increases microorganism clustering: The implications for microbiological risk assessment. Journal of Water Supply: Research and Technology - *Aqua* **46**,117-126.

Hall, T. (1998) *A Guide to Water Treatment Practices*. WRc publication No 854.

Hoff, J.C. and Geldreich, E.E. (1981) Comparison of the biocidal efficiency of alternative disinfectants. *Journal of the American Water Works Association* **73**, 40-44.

Hunt, N.K. and Mariñas, B.J. (1999) Inactivation of *Escherichia coli* with ozone: chemical and inactivation kinetics. *Water Research* **33**, 2633-2641.

IWSA (1998) Proceedings of IWSA International Conference. Drinking water distribution with or without disinfectant residual. *Water Supply* **16** (3/4).

Jacangelo, J.G., Adham, S.S. and Laîné, J.-M. (1995) Mechanism of *Cryptosporidium*, *Giardia*, and MS2 virus removal by MF and UF. *Journal of the American Water Works Association* **87**, 107-121.

Jagger, J.H. (1967) *Introduction to Research in UV Photobiology*. Prentice-Hall, Inc., Englewood Cliffs, NJ, United States.

Kaiser, H.P., Von Gunten, U. and Elovitz, M. (2000) Die Bewertung von Ozonreaktoren. *Gas, Wasser, Abwasser* **80**, 50-61.

LeChevallier, M.W. and Au, K.K. (2002) Water treatment for microbial control: A review document. World Health Organization.

Mechsner, K., Fleischmann, T., Mason, C.A. and Hamer, G. (1991) UV disinfection: Short term inactivation and revival. *Water Science and Technology* **24**, 339-342.

Reed, R.H. (1997) Solar inactivation of faecal bacteria in water: the critical role of oxygen. *Letters in Applied Microbiology* **24**, 276-280.

Rubin, L.B., Burchuladze, T.G. and Fraikin, G.Y. (1982) Two-photon inactivation, photoreactivation and photoprotection in yeast cells irradiated by 266 nm-laser radiation. *Photochemistry and Photobiology* **35**, 789-791.

Snozzi, M., Haas, R., Leuker, G., Kolch, A. and Bergman, R. (1999) Prüfung und Zertifizierung von UV-Anlagen. *Gas, Wasser, Abwasser* **79**, 380-385.

Snozzi, M. (2000) New concepts and methods for the evaluation of the microbial quality of drinking water. *Mitt. Lebensm. Hyg.* **91**, 44-52.

Sobsey, M.D. (1989) Inactivation of health-related microorganisms in water by disinfection processes. *Water Science and Technology* **21**(3),179-195.

Sobsey, M.D. (2002) Household water treatment and storage as appropriate technology for the developing world. World Health Organization

Sommer, R. and Cabaj, A. (1993) Evaluation of the efficiency of a UV plant for drinking water disinfection. *Water Science and Technology* **27**, 357–362.

Sommer, B., Mariño, A., Solarte, Y., Salas, M.L., Dierolf, C., Valiente, C., Mora, D., Rechsteiner, R., Stters, P., Wirojanagud, W., Ajarmeh, H., Al-Hassan, A. and Wegelin, M. (1997) SODIS – an emerging water treatment process. *Journal of Water Supply: Research and Technology – Aqua* **46**(3), 127-137.

Taylor, J.S. and Wiesner, M. (1999) Membranes. In: *Water Quality and Treatment*. R.D. Letterman (Ed.) McGraw Hill, Inc., New York. p. 11.1-11.71.

USEPA (1989) *Guidance Manual for Compliance with the Filtration and Disinfection Requirements for Public Water Systems Using Surface Water Sources*, US Environmental Protection Agency, Washington, D.C.

USEPA (1999) *Guidance Manual Alternative Disinfectants and Oxidants*. EPA 815-R-99-014. US Environmental Protection Agency, Washington, D.C.

Wegelin, M., Canonica, S., Mechsner, K., Pesaro, F. and Metzler, A. (1994) Solar water disinfection: Scope of the process and analysis of radiation experiments. *Journal of Water Supply: Research and Technology - Aqua* **43**(3), 154-169.

White, G.C. (1999) *Handbook of Chlorination and Alternative Disinfectants*. John Wiley and Sons, Inc., New York.

*Chapter 6*

## MONITORING THE QUALITY OF DRINKING WATER DURING STORAGE AND DISTRIBUTION

*W. Robertson, G. Stanfield, G. Howard and J. Bartram*

## 6.1 Introduction

Following abstraction and treatment water becomes a vulnerable and perishable product. It is vulnerable in that the integrity of systems used for the storage and distribution of water can be damaged and contamination through ingress can occur. It is perishable in that its microbial quality can deteriorate due to the bacteria remaining after treatment growing on the residual nutrient in the water. Water can therefore be regarded as having a finite life.

The rate at which quality deteriorates can be controlled by adding a preservative (disinfectant residual), or by the use of advanced treatment to remove as much biodegradable organic carbon as possible from the water. However, these measures will be to no avail, if the distribution or storage-systems are contaminated or permit ingress. To get the maximum degree of protection the product (water) and the packaging (tanks and pipes) must be clean and intact.

The importance of quality changes in distribution is based upon evidence concerning the frequency and extent of known quality changes and their impact upon human health. Thus, for example, a significant proportion of recognised piped drinking water-related disease outbreaks are related to quality deterioration in distribution (Ainsworth, 2002; Craun and Calderon, 2001). However, most water supply world-wide is unreliable because of, for example, intermittence and so household management is common. This may involve only plumbed in household storage tanks or alternatives, which involve extensive manual handling. Because the contamination that arises is local in character it is unlikely to give rise to detectable outbreaks of disease but significant evidence exists that quality changes in such circumstances may be extreme and respond to the extent of handling (Quick *et al.*, 1999).

## 6.2    Piped distribution systems

Piped distribution systems for drinking water are as important to the quality and safety of drinking water as the treatment itself. Water entering the distribution system must be microbiologically safe and ideally should be biologically stable. The distribution system itself must provide a secure barrier to post-treatment contamination as the water is transported to the user. Residual disinfection will provide partial protection against recontamination, but may also act to mask the presence of such contamination. On a global scale, however, mishandling within the home is likely to be the most significant source of fouling. This section describes the sources of contamination progressing from treated water leaving the treatment plant, to deterioration of water quality during storage and distribution. Non-piped systems are dealt with in a separate section.

### *6.2.1    Inadequately treated water entering the distribution system*

Sporadic occurrences of source water of high turbidity that overwhelm the treatment train, or microorganism breakthrough for example resulting from sub-optimal filtration following filter backwashing, can introduce enteric pathogens into the distribution system. These may be in sufficient numbers to cause detectable cases of gastrointestinal illness in the population served (MacKenzie et al., 1994). It should be recalled that the frequency of such events in developed and developing countries is likely to be far greater than that implied by the number of outbreaks presently detected (see Chapter 7). Such occurrences are often rainfall driven and may be inter-related. For example exceptionally high pathogen loadings may arise from surface runoff following rainfall at the same time that overall filter efficiency decreases and the need for backwashing increases because of the concurrent high turbidity loadings (see Chapters 4 and 5).

### 6.2.2    *Integrity of the distribution system*

Disease outbreaks have been traced to cross connections in spite of the fact that water leaving the plant was deemed to be safe. In general, there are four types of cross connections:

- **Infiltration**. In this situation contaminated sub-surface water is drawn into the distribution system. In order for this to happen three conditions must be in place. First, contaminated water must be present in the sub-surface material surrounding the distribution system, possibly from a leaking sanitary, storm or combined sewer. Secondly, there must be an adjacent low-pressure zone within the system. These zones can arise through high water usage resulting from fire fighting or other peak demands, decreased flow arising from restrictions in the system, pump failures or intermittent operation of the treatment plant. More recent evidence points to the role of pressure surges, in otherwise properly operated systems, in creating transient low pressure that may lead to the ingress of contaminants (LeChevallier, 1999). Thirdly, there must be a route for contaminated water to enter the system. This can occur through pinholes caused by corrosion, cracks or outright breaks or leaking joints in the wall of the mains. If these three situations occur simultaneously then it is likely that contaminated water will enter the distribution system. Leakage rates are typically high, with even well operated systems experiencing rates of 10 – 20% (LeChevallier, 1999; WHO and UNICEF, 2000). This is likely to be associated with a relatively large number of points of leakage and therefore present an increased risk of intrusion of pathogenic organisms.

- **Back siphonage**. In this situation faecally contaminated surface water is drawn into the distribution system or storage reservoir through a back flow mechanism. In order for this to take place two conditions must occur simultaneously. Firstly, there must be a reduction in line pressure as described above. Secondly, there must be a physical link between contaminated water and the storage and distribution system. Open taps connected to hoses that are submerged in pools of water may provide this link. Back flow preventers are available to stop such occurrences. Plumbing codes should require air gaps between taps and the brim of receptacles. The general conclusion stresses the importance of good domestic plumbing practice.

- **Open drinking water storage reservoirs**. Microbial contamination can also be introduced into the distribution system through open treated-water storage reservoirs (Geldreich, 1996). For example, waterborne outbreaks have occurred in communities where birds contaminated the water either because the reservoir was uncovered or because they gained access to the reservoir through unscreened roof vents. Uncovered reservoirs can also permit the growth of toxin-forming cyanobacteria.

- **Line construction and repair**. When existing mains are repaired or replaced or when new water mains are installed strict protocols involving disinfection and flushing must be followed to prevent the introduction of contaminated soil or debris into the system (*e.g.* AWWA, 1986). These protocols generally cover six areas of concern:

    1. Protection of pipe sections at the site.

    2. Restriction on the types of joint sealing materials used.

    3. Preliminary flushing of pipe sections.

    4. Disinfection of pipe sections.

    5. Final flushing of pipe sections.

    6. Bacteriological testing to confirm disinfection.

If these procedures are not adhered to pathogens may gain access to the system. Construction materials, such as wood embedded in pipe sections, have also been identified as a source of microbial contamination and provide an adequate supply of nutrients to support bacterial re-growth (Martin *et al.*, 1982).

In each scenario, if the contaminated water contains enteric pathogens then it is likely that consumers will be exposed to them. Even where disinfectant residuals are employed to limit microbial occurrence it may be inadequate to overcome the contamination or may be ineffective against some or all of the pathogen types introduced. As a result, pathogens may occur in concentrations that could lead to infection and illness. In many developing countries drinking water is supplied intermittently either as a cost-saving measure or because of water shortages. Under such conditions the resulting low water pressure will allow the ingress of contaminated water into the system through breaks, cracks, joints and pinholes in the walls of the system.

In the context of monitoring it should be noted that the use of residual disinfectants to control such problems may lead to an inaccurate estimation of their occurrence. This is because the microbial indicator parameters most frequently used for monitoring are among the most sensitive of the organisms of interest, leading to a potential situation where the indicator parameter is absent but pathogens may be present. Thus, neither the occurrence nor the severity of contamination will be properly identified.

### 6.2.3    *Microbial re-growth in the distribution system*

Even though treated drinking water may be free of faecal indicator organisms and detectable enteric pathogens and therefore present a tolerable level of risk of enteric illness, drinking water entering the distribution system may contain free-living amoebae and environmental strains of various bacterial species, often referred to as heterotrophic bacteria. Under favourable conditions amoebae and heterotrophs will colonise a distribution system and form biofilms.

Many environmental strains of coliform bacteria such as *Citrobacter*, *Enterobacter* and *Klebsiella* may also colonise distribution systems (Martin *et al.*, 1982). However, it is generally agreed that water temperatures and nutrient concentrations are not elevated enough to support the growth of *E. coli* (or enteric pathogenic bacteria) in biofilms (Geldreich and LeChevallier, 1999). Thus the presence of *E. coli* should be considered as evidence of recent faecal contamination of drinking water.

Many species of bacteria and free-living amoebae can occur in biofilms within distribution systems including certain opportunistic pathogens. There is inadequate evidence at present to implicate the occurrence of these microorganisms from biofilms (excepting for example, *Legionella* or *Mycobacterium*) with adverse health effects in humans with the possible exception of immuno-compromised population groups (reviewed by Geldreich, 1996).

## 6.3    Non-piped distribution systems

Most of the population of the globe obtain their water through non-piped systems and of the remainder the majority are supplied through systems that require some form of storage and/or handling before use, thereby increasing the possibility of contamination. Even where a reliable piped supply is the norm, occasional interruptions occur (*e.g.* extreme weather events that cause major line breaks and force adoption of household handling). From a public health viewpoint it is therefore essential to respond to contamination that occurs through the chain of supply up to the point of use and to consider all forms of water supply used by the population.

Point sources of water such as tubewells, dug wells and protected springs represent a very significant proportion of the 'improved' water supplies provided to communities in developing countries (WHO and UNICEF, 2000). Such supplies are very common in rural areas and may also represent a very significant proportion of the water supplies available and used for domestic purposes (including drinking) by low-income urban populations (WHO and UNICEF, 2000; Howard *et al.*, 1999; Ahmed and Hossain, 1997). The quality of such sources is often very variable and they frequently show gross faecal contamination, particularly during wet seasons (Wright, 1986; Barrett *et al.*, 2000). The public health consequences of consumption of contaminated water from point sources can be severe both in relation to endemic and epidemic disease (Pedley and Howard, 1997). The control of the quality of drinking water from such sources is important in reducing public health risks, but requires greater emphasis on support to community management in order to improve operation and maintenance, and also significant user education.

As point water sources are often found in areas that also rely on on-site sanitation, sub-surface leaching is frequently identified as being a principal source of contamination (Melian *et al.*, 1999; Rahman, 1996). However, there is increasing evidence that deterioration in sanitary completion measures and preferential flow paths are more important in causing microbial contamination (Howard *et al.*, 2001a; Gelinas *et al.*, 1996; Rojas *et al.*, 1995). Different technologies have different vulnerability to contamination depending, largely, on the depth of abstraction (ARGOSS, 2001).

Many people world-wide rely on water supplied by tankers or other forms of vendor supplies (Whittington *et al.*, 1991). The water in such supplies may come from hydrants connected to utility supplies or may be drawn from alternative sources. In many cases, the consumer will not be aware of the source of the water and there may be significant concerns about the quality of water (Lloyd *et al.*, 1991). Where the water is supplied from hydrants via utility piped water supplies, basic good hygiene practices, such as regular cleaning and disinfection of the tank and sanitary storage of the connecting hoses is usually adequate to maintain quality. Studies in Ghana, showed that the quality of water in tanker trucks collecting water from utility hydrants was the same as the piped water in Kumasi and only marginally less good in the capital Accra (Jabulo, *pers. comm*). In the Ghana study and also studies in Uganda (Howard and Luyima, 2000), however, it has been shown that the quality of water supplied by other vendors, such as those selling from jerry cans or small fixed tanks, is much worse and represents highly degraded water quality.

In non-piped systems drinking water is typically either carried to the home or is transported by truck or cart. The contamination of the water, usually as a result of poor hygiene is common. Source contamination can then be augmented at each point of handling through collection and transport (VanDerslice and Briscoe, 1993). However, because such contamination is by definition local in nature, it is unlikely to lead to large-scale single source outbreaks and therefore unlikely to be detected or reasonably estimated.

Poor hygiene in the home is another, potentially significant source of drinking water contamination. In many cases, contamination significantly increases from source to household (WHO, 1997). This is of particular concern in communities without reliable piped water to within dwellings which, therefore, rely upon water storage containers. Evidence is accumulating for both the impact of such contamination on human health and for the effectiveness of interventions at this level in protecting and improving human health (Semenza *et al.*, 1998). Such interventions include the use of household water disinfection technologies, including chlorination and solar disinfection (see Chapter 5; Sobsey, 2002). They also include hygiene education interventions geared to the maintenance of water quality during transport and storage. Such interventions may include the testing of water and participatory approaches to health education, which, have proven successful in both rural and urban areas (Breslin, 2000).

The selection of whether water treatment or hygiene education interventions are most appropriate depends in part on the quality of the source water. If it is good, then greater effort may be placed on promoting safe handling, whereas poor source water quality greater effort may be placed on household treatment. Where household disinfection is promoted, it is essential that water quality programmes provide educational information on disinfecting drinking water in the home and maintaining the quality and safety of the water following disinfection. There is considerable evidence to suggest that this approach is effective at reducing contamination and the spread of enteric illness (Sobsey, 2002; Forget and Sanchez-Bain, 1999; Quick *et al.*, 1999; Reiff *et al.*, 1996).

## 6.4    Approaches to sampling in piped distribution systems

Strategies for monitoring the microbiological quality of water in supply must be designed to allow the best possible chance of detecting reductions in microbial quality (Ainsworth *et al.*, 2002). It is vital, therefore, to ensure that the samples taken are representative of the quality in the distribution network.

In formulating a strategy the factors outlined in the following subsections need to be considered (ISO, 1991).

### 6.4.1    Choice of indicator parameter

Traditionally, microbial indicators of faecal pollution such as total coliforms and *E. coli* have been the determinants used. Most national and international microbial standards for water in supply are expressed in terms of these organisms. Measurement of the numbers of heterotrophic bacteria provide further valuable, but often under-utilised, additional information. Secondary microbial parameters, for example, enterococci and clostridia, being more resistant to disinfection, may also be used to try and gain better insight into the source of the contamination or where particular sources of contamination are suspected or known to cause problems. These secondary parameters are most often used, therefore, as part of investigations where a failure of quality, in terms of the detection of coliform bacteria, has been detected. The practice of application of a residual disinfectant such as chlorine has a significant impact upon the adequacy of alternative indicators as outlined above. Since the most common microbial indicators of faecal pollution (*E. coli* or thermotolerant coliforms and total coliforms) are very sensitive to chlorine their detection implies the likelihood of either recent or substantial faecal contamination with attendant health risks. Nevertheless, because coliform bacteria are so sensitive to chlorine their absence provides no guarantee that pathogens such as enteric protozoa and viruses, which are more resistant to chlorine, are also absent.

In any faecal contamination event microorganisms, whether introduced by inadequate treatment or post-treatment contamination, are not distributed evenly throughout the distribution system but are typically clumped (Gale, 1996a). In these situations the probability of detecting faecal indicator bacteria in the relatively few samples collected from the distribution system during routine sampling is substantially reduced. Hence quantitative measurements of contamination, such as membrane filtration (MF) or most probable number (MPN) methods, may provide a poor estimate of the overall density of indicator bacteria. Frequency-of-occurrence monitoring, based on the Presence/Absence (P/A) test, can provide a better estimate of water quality in general (Pipes and Christian, 1984; Clark, 1980). In this method the frequency of positive samples detected during routine sampling during a reporting period is compared with acceptable frequencies of positive samples specified in the applicable water quality standards. Where such standards are based upon *E. coli* or thermotolerant coliforms, their confirmed presence usually leads to an immediate boil water advisory and corrective actions. Standards based upon the presence of total coliforms specify an acceptable frequency of positive samples (for example 5%) before resampling and possible corrective actions are required. Comparative studies of the P/A and MF methods demonstrate that the P/A method can maximise the detection of faecal indicator bacteria (Clark, 1990; Geldreich, 1996). It also allows more samples to be analysed within a reporting period because the test is simpler, faster and less expensive than the quantitative methods noted above (see Chapter 2.2.9). Commercial P/A kits for faecal indicator bacteria are available.

More conservative microbial parameters (*i.e.* parameters of quality deterioration that may be detected before actual faecal contamination occurs) are logically preferred but most present practical/logistical problems. The most frequently used alternative for distribution systems is presently the heterotrophic plate count. Whilst taxonomically imprecise this is relatively widely available and applicable. Its value is principally associated with changes and trends in counts retrieved from a given system rather than comparison with numeric reference values and it therefore requires relatively dense sampling networks (in time and space) to provide useful information. Measuring the numbers of heterotrophic bacteria present in a supply can be a useful indicator of changes such as increased re-growth potential, increased biofilm activity, extended retention times or stagnation and a breakdown of integrity of the system. Following trends in the numbers of heterotrophs may therefore be useful in prompting, for example, a programme of mains flushing or cleaning.

Non-microbial parameters may also be suitable for this purpose and also require comparison on changes and trends and therefore relatively dense sampling networks. The most frequently used examples include conductivity/total dissolved solids, turbidity and chlorine residual (where chlorine is applied). Where chlorine residuals are used, measurement of this residual can often be a more rapid indication of problems than microbial parameters. A sudden disappearance of an otherwise stable residual can indicate ingress of contamination with a high organic loading. Alternatively, difficulties in maintaining residuals at points in a distribution system or a gradual disappearance of residual may indicate that the water or pipe work has a high oxidant demand due to re-growth of bacteria in the water or biofilm growth.

Routine monitoring of distributed water for particular groups of bacteria or specific pathogens is rarely considered worthwhile or necessary. However, in the Netherlands guidelines for the numbers of aeromonads in final waters and in distribution have been issued to act as an indication that better maintenance of sand filters or that better removal of methane from anaerobic groundwater sources is required. In this context aeromonads are acting as a sensitive indicator of the potential for re-growth to occur within the network (see Chapter 2.2.8).

### 6.4.2    Location of sampling points

The two principal considerations in the location of sampling points are whether fixed or variable locations (or a mixture of these) are to be employed, and whether sampling is to occur from normal access points (such as consumer taps), from dedicated but otherwise normal locations or from structures dedicated exclusively for this purpose. The purpose of monitoring and especially whether in response to statutory requirements or for public health investigatory purposes will have a significant impact on the choices made.

Sampling locations should be chosen to provide a means of characterising water quality in all parts of the system. For this reason a supply system may be divided into a series of zones on the basis of geographical area, the size of the population served or specific areas of the pipe network. Sampling points are then identified within each zone to try and ensure that representative samples are taken. Sampling points will be chosen for two reasons. One will be to satisfy statutory responsibilities and the other for strategic or operational purposes. In the latter, a water supplier may for example be trying to obtain more information about an area that has, in the past, yielded frequent coliform failures or elevated counts of heterotrophs.

Fixed sampling points are frequently used, and may be chosen because of ease of access. Often these points are located within public buildings, or in the premises of public services, such as fire-stations. The use of fixed sampling points alone can be regarded as unsatisfactory, since they may not give a representative view of what is happening in all parts of the distribution system or zones. To overcome this, additional samples may be taken from other outlets chosen at random and these usually include the mains tap in the houses of consumers.

The use of fixed or random sampling points is a topic of much continuing debate. Geldreich (1996) reported that in a survey of 1 796 water supply utilities in the USA, about one third used only fixed sampling points, with 50% using a combination of random and fixed points. He recommended that some sampling locations be varied, so that all sections of the network are monitored over time. Burlingame and O'Donnell (1993) argue that the use of random taps increases the detection of water contaminated within households (which is outside the regulatory framework), since the water supplier has no control over consumers' premises. Similarly Dufresne et al. (1997) have demonstrated that the number of positives and the number of samples collected can be reduced by selecting sampling points that were protected and well maintained. Similarly, a task group, appointed by United Kingdom Water Industry (1995), concluded that there were advantages in adopting a wholly fixed-point system. These included the control and maintenance of the taps, greater compatibility of data and reduced cost. Other studies have indicated that the type of tap, particularly mixer taps, and material of construction may also influence water quality. These factors need to be considered in the design of a monitoring strategy. If random taps are to be used, it may be that identification of premises with suitable taps may be a more important consideration than achieving true randomness (Anon, 1994).

For statutory purposes, the location and number of sampling points may be stipulated within the appropriate regulations of that country. For example, the UK Water Supply Regulations stipulate that at least 50% of distribution samples must be taken from random locations (HMSO, 1989). For strategic monitoring, locations will be selected to gain the best practicable information about the area of the system being investigated. For this purpose, depending on the laboratory resources available, sampling locations may be more numerous than required to satisfy the regulations. Regardless of the reason for the monitoring it will have the common objective of providing sufficient information to allow the water quality in all service areas to be characterised.

### 6.4.3    *Frequency of sampling*

There is no absolute rationale for the frequency of sampling and adopted frequencies reflect the need to balance the benefits of better information arising from greater frequency with increasing costs (and decreasing returns) as overall frequency increases. The guidelines and regulations of individual countries generally prescribe the minimum sampling frequency to be achieved to meet statutory requirements. These sampling frequencies are usually based on the population served by the network/zone or, less frequently, on the volume of water supplied. For example in France, a minimum of 24 samples each year must be taken from distribution systems serving 10 000 – 20 000 inhabitants. In the Netherlands the frequency is similar, while the UK Regulations prescribe a more intensive programme of 48 samples per annum. In Germany the prescribed frequency is similar to that set for treatment works at one sample per 15 000 cubic metres of water supplied.

Although jurisdictions prescribe minimum sampling frequencies it is often necessary to collect additional samples to improve the overall picture of water quality in the distribution system. Supplementary samples should be collected at locations which, from historical data, are known to experience problems. Sampling frequency should also be increased following remedial actions, for example in response to boil water orders, or following interruptions in supply. In most cases a failure in terms of *E. coli* or thermotolerant coliform detection will initiate re-sampling, which may be followed by intensive investigations to identify the source and extent of the contamination. While re-sampling is commonly required the rationale underlying this (as a procedure prior to further investigation and action) is unclear. Given the known temporal and spatial variability of microbial water quality it is logical that re-sampling would often fail to detect continued contamination without indicating that the cause of the original contamination had been limited or controlled.

Some water suppliers carry out much more sampling within the distribution network than required to satisfy the regulations. Strategic monitoring programmes will be carried out to investigate specific problems such as high incidence of taste and odour events or the need for mains replacement or renovation. Each distribution system will have its own unique monitoring needs and the monitoring programme should be designed to address these using the available analytical resources. The impact of the timing of sampling should be recalled in interpreting results. Thus, for example, most sampling will be undertaken within the normal working week and may not detect changes arising from abnormal patterns of demand (such as football cup finals).

### 6.4.4    Volume of sample

The volume of sample collected must be sufficient to allow analysis at the limits of detection stipulated in regulations. For coliform bacteria this is 100 ml. Heterotrophic plate counts are usually expressed on a per ml basis. Given that samples may have to be re-analysed, the volume of sample will therefore generally be between 250 and 500 ml. In the Netherlands, coliform and *E. coli* counts are expressed per 250 ml and so sample volumes will be greater to satisfy these requirements.

Demonstrating the absence of indicator microorganisms in 100 ml volumes, although adequate for statutory requirements, does not provide information about how close to compliance water quality is. For this reason some water suppliers routinely, or as part of special investigative studies, may collect and analyse sample volumes of between 1 and 10 litres. Volumes of this magnitude can be analysed by small modifications to existing analytical methods or by multiple analyses of 100 ml to 1 l volumes.

If microorganisms were distributed randomly in distribution systems, the use of the Poisson distribution would be appropriate in calculating the confidence limits for occurrence of indicator bacteria (Haas and Heller, 1986). However, it has been suggested by several workers (*e.g.* Pipes *et al.*, 1977;

190

Gale, 1996a) that the distribution of bacteria tends not to fit this pattern, instead they occur in clusters within a bulk supply. Gale (1996b) analysed the statistical distribution of confirmed coliform densities and total heterotrophic bacteria densities, using monitoring data from eight UK water companies. It was found that the log value of the bacterial density varied in an approximate linear fashion when plotted against the percentage of the data that was less than this value: this is known as a log-normal distribution. Extrapolation of the log-normal distribution of confirmed coliforms (1-9% of all samples) implied that where 100 ml samples were recorded as 0/100 ml, the actual concentration may have been as low as $< 1/10^7$ litres or as high as 1/litre. Similar clustering of a small proportion of high counts could be caused by sporadic contamination of the distribution system through, say, floc carryover and filter breakthrough (Gale, 1996b). Gale (1996b) concludes that ideally a model of the risk of exposure to pathogens would take account of the density of the organism in raw water, the removal efficiency by subsequent treatment processes, the variation in tendency to cluster and an estimate of water consumption.

In addition to providing some indication as to the true numbers of microbial indicators of faecal pollution in distribution, analysis of large volume samples increase the chance of detecting clusters or pulses of bacterial contamination particularly if taken over a period of time such as 30-60 minutes. An alternative to large volume grab samples is the collection of a composite sample. This is composed of a series of small volume samples collected at time intervals that are bulked together to form one large sample. Depending on available analytical resources analysis of individual samples would provide an indication of the bacterial distribution. It can be difficult to arrange for large volume or composite samples to be collected at randomly selected locations such as consumers' taps, and it is best restricted to fixed sampling points.

### 6.4.5    Sample collection

Sampling for microbiological analyses requires care, with observation of the general principles of aseptic technique. All equipment used for sampling should be thoroughly cleaned and preferably sterilised before use. The sample containers should be sterile with wide mouths that are shielded from contamination. They should contain sufficient sterile reagent to neutralise any residual disinfectant in the water and samplers should be trained to ensure that thorough mixing occurs immediately after sampling.

For some purposes and especially in relation to the performance of the supply agency the sampling procedure used has to ensure that water from the main is collected and not the stagnant water in the supply pipe. For other

191

purposes (especially in relation to public health investigation) interest may focus on the water that has gone through typical processes of deterioration, as this is what consumers will be exposed to. Collecting mains water is usually achieved by running water to waste for a period of two-three minutes (Anon, 1994; APHA, AWWA, WEF, 1998), but given the differing lengths of service pipes this may not always be adequate. Other methods, such as running to waste until a constant temperature is reached (Anon, 1994), particle counts have stabilised (Burlingame and Choi, 1998) or (if appropriate) a chlorine residual is detected would appear to be more certain, but rather time consuming. After reviewing the results of studies on flushing Prévost et al. (1997) suggest a period of five-ten minutes may be necessary. However, the consumers whose taps are being used do not like these methods because they are perceived as being wasteful (or costly, if on a metered supply). The importance of this step in the sampling procedure cannot be over emphasised. Since water in the service pipe may reach relatively high temperatures, heterotrophs may grow to higher densities than in the water main (Geldreich, 1996). In addition, the high surface area to volume ratio of service pipes encourages the decay of chlorine residuals, again allowing a greater potential for bacterial growth. Without an adequate period of flushing a representative sample will not be obtained. However, not all workers agree with this. Prevost et al. (1997) demonstrated that although numbers of bacteria increased with distance from the treatment works, stagnation within the service pipe did not influence numbers appreciably. Similarly Kerneis et al. (1995) found that residence time had little influence on the heterotrophic plate count.

The introduction of new designs of taps (mixer), the use of plastics in construction and the provision of inserts as anti-splash devices has caused much discussion on the need for, and the best method of, sterilising taps prior to sampling. The traditional method of sterilisation is to heat the outside of the tap, in the direction of spout to base. Although this method is satisfactory where metal taps are used at premises controlled by the water supplier, they may be considered too hazardous at consumers premises. For sampling at these locations heat has almost completely been replaced by chemical methods of disinfection. Sodium hypochlorite is the most widely recommended chemical disinfectant. It is effective against a wide range of bacteria, including spores, and viruses. Only fresh solutions should be used as the concentration of chlorine decays with time. The consensus opinion on concentration is to use a 10% solution of household sodium hypochlorite for swabbing, or a 1% solution where a spray is used. It is essential that tap surfaces are clean before the use of a chlorine-based disinfectant since they react readily with all types of organic matter. In some cases isopropanol is used for tap disinfection as it is effective against some of the more resistant bacteria, such as mycobacteria, it is water miscible and evaporates at room temperature, leaving no residue. No particular precautions other than simple tap flushing, have to be taken to ensure that the disinfectant does not contaminate the sample. In common with most chemical disinfectants, it precipitates proteins that may then form a protective layer for bacteria. The tap body and spout must therefore be clean before application. Isopropanol has the disadvantages that prolonged contact may result in skin irritation, swelling of rubber and hardening of some plastics. Burlingame and Choi (1998) suggest that the concentration used should be at least 70%, but a slightly lower concentration 60-70% is said by Gardner and Peel (1986) to be more effective.

## 6.5    Approaches to sampling in non-piped systems

In many areas where communities rely on non-piped supplies, resources to carry out routine bacteriological monitoring are very limited. This is often further complicated as most small water supplies are community-managed and therefore it is difficult to ensure that the "supplier" is able undertake an expected level of quality control monitoring and therefore there is greater reliance on external verification. Furthermore, enforcement of action is usually difficult as community-managed systems are often difficult to regulate directly and this becomes impossible when water hygiene is evaluated (Howard *et al.*, 2001a). It is possible, however, to undertake programmes as described below. Given that monitoring microbial quality may be expensive, when considering the development of sampling programmes for point sources, it is essential that the monitoring programme is designed to meet a clearly defined management need and that the data derived will result in useful information (Bartram, 1999; Adriaanse, 1997; Ongley, 1998).

As for all water supplies, the use of sanitary inspection is extremely important as a means of assessing long-term risks and in analysing the causes of contamination when this is found. Sanitary inspections are visual assessments of the infrastructure and environment surrounding a water supply taking into account the condition, devices, and practices in the water supply system that pose an actual or potential danger to the health and well-being of the consumers (WHO, 1997). The most effective way to undertake sanitary inspections is to use a semi-quantitative standardised approach using logical questions and a simple scoring system. There exist a number of example forms that can be used (WHO, 1997). Sanitary inspections are complementary to water quality analysis and there is an increase in the power of subsequent analysis when both types of data are available.

### 6.5.1    Selection of methods and indicator organisms

For communities where point sources are commonly used, the use of on-site testing kits is often recommended given the large distances between the sources and laboratories and the very significant problems with sample deterioration (Bartram and Ballance, 1996). There appears to be no significant difference in the reliability of results obtained from such kits in comparison to laboratory testing providing the staff using them are properly trained and maintain an aseptic technique. Kits using MF, MPN and P/A methods are available.

Where bacteriological testing is undertaken using test kits, it is likely that *E. coli* or thermotolerant coliforms will be the indicator parameters analysed. However, for shallow groundwater, the use of faecal streptococci may provide more reliable results given their greater environmental resistance and because there is evidence that thermotolerant coliforms may multiply in nutrient enriched environmental waters (WHO, 1996; Byappanahalli and Fujioka, 1998). Studies in Uganda using sorbitol-fermenting bifido-bacteria, which are unique to human faeces showed a stronger relationship with faecal streptococci than thermotolerant coliforms (Howard *et al.*, 2001b). Most kits for thermotolerant coliform analysis can also be used for faecal streptococci analysis, although they must be able to sustain 48 hour incubation times.

### 6.5.2    Water quality sampling approaches

The degree to which sampling of point water sources is developed is dependent largely on the objectives of the monitoring programme being implemented. Very few countries have fully developed programmes of rural water supply testing. Indeed, although such approaches do provide an overall indication of water quality and its variation, it does not necessarily yield data from which management decisions can be easily made.

There are two main approaches (years-group approach and longitudinal study sample) that can be recommended for sampling of point water sources, which are outlined below. In addition, a brief section is included regarding assessment-only approaches, which are not recommended. The recommended approaches work on a slightly different basis and provide somewhat different information. Common to both approaches, however, is that a sample of water sources are tested during each sampling round.

The first stage in both approaches is to compile an inventory of all the water sources to be included in the monitoring programme (Lloyd and Bartram, 1991). A second key element in both approaches is to ensure that testing is timed to coincide with those times when water quality may be most threatened, usually during the wet season. However, is should be noted that some shallow groundwater systems show very rapid response to rainfall and this may need to be considered when designing the sampling programme (Barrett *et al.*, 2000). Samples should be taken from the principal outlet – handpump, spring outlet or bucket used to take water from a well.

### 6.5.2.1   *Year-groups approach*

In this approach, all the water supplies are assigned to a particular "year-group" and a rolling programme of visits developed (Bartram, 1999). All the sources listed in each "year-group" would be visited, with the aim that all supplies would receive repeated visits over a time frame of between two to five years. On each visit, detailed surveys would be undertaken including sanitary inspection, water quality analysis and household visits made. Key to this approach is to ensure that stratified or cluster sampling approaches are adopted to ensure that supplies in different parts of the country were included within each 'year group'. If this is not done, then the results from particular "year groups" may provide a distorted picture of the microbiological quality of water, as it may be biased because of technology type, hydrogeology or pollution loading specific to particular areas. The "year-group" approach provides the monitoring body a greater breadth of information as it attempts to cover all water sources. The data may also be used to identify actions required to improve water quality, although the restricted numbers of samples at sources may mean that it is more difficult to develop a full understanding of water quality variation.

### 6.5.2.2   *Longitudinal study sample*

In this approach, a representative sample of water supplies is visited on a regular basis. It is therefore different from the above approach in that it does not attempt to provide data on all sources within the country, but rather addresses developing a more in-depth understanding of the variation in quality of different types of water sources and different parts of the country. This can be used to assess whether microbial quality and risks vary over time and provides useful insight into how effective community operation and maintenance is and what improvements are required both in terms of training and design. The longitudinal study approach aim is to use the data from the sample to inform overall implementation and management of the quality of water by ensuring that the most important factors in causing microbial contamination are identified and preventative and remedial actions developed. These apply beyond the sample of water sources included within the clusters and can form the basis of a broader national strategy for water quality improvement.

### 6.5.2.3    *Assessment only approaches*

In some countries, sampling of the water sources only occurs during the source selection stage and there are often particular levels of microbial contamination (usually defined on the basis of thermotolerant coliforms or more rarely *E.coli*) above which the source should not be used. A common value used is 50/100ml. However, such approaches have very limited use as the results of a single test may not provide a realistic estimate of the microbial quality of the water (particularly as this would not typically be done during worst case situations) and the use of an arbitrary figure of 50/100ml is unlikely to be meaningful in terms of health risk. Clearly, such approaches also fail to provide any indication of whether the protection works undertaken have been effective in reducing pollution and the designs used have rarely, if ever, been properly evaluated in terms of their ability to reduce contamination as expressed through log-reductions in bacterial densities.

Other countries either complement or replace the testing of sources during selection with a test on commissioning. This has the advantage that it provides a better indication of the quality of the supply, but again may not reflect seasonal variation in quality. Such approaches will also fail to address the greater concerns of water quality deterioration over time in supplies where operation and maintenance is weak.

### 6.5.4    *Tanker trucks and vendors*

These are treated separately as vending is a commercial practice and therefore more amenable to direct control than when water supplies are community-managed. Where trucks collect water from water supply utility hydrants, the routine testing of the quality of water in the trucks and at the hydrants should form part of the routine distribution testing by the water supplier. In the first instance an assessment should be carried out covering a sample of filling stations and tankers, followed by a lower-intensity routine monitoring programme when the day of sampling, selection of hydrants and selection of trucks vary in order to ensure to avoid biased results. This can be linked to codes of sanitary practice both by the water supplier and the tanker trucks (Lloyd *et al.*, 1991).

Where vendor supplies are not taken from utility piped supplies, routine testing and sanitary inspection is still desirable, although the type of vending may dictate the ease with which this is implemented. For tanker trucks, a rolling programme of random testing can be initiated, although it may be difficult to identify source waters. For very small vendors, it may not be possible to undertake routine monitoring, although occasional assessments would be worthwhile. In both cases, it is critical that the monitoring will lead to some form of action, whether this is through working with vendors to improve practices, regulating vending practices or banning all vending.

### 6.5.5    *Household water*

The testing of water stored in households is important to ascertain the quality of water actually being consumed. This is important because post-source deterioration in quality may have occurred and therefore good quality water at source may have become severely degraded by the time it is consumed and remedial actions (for instance hygiene education programme) may be required. The testing of household water is therefore an important component of an integrated risk-based approach to water quality.

Household samples should be taken from the drinking water storage vessel used by the family and containers used for collecting and transporting the water. A water chain can be tracked from source to storage with samples taken from source, collection vessel and water storage container. This often provides useful information regarding where and what type of interventions (educational, technical) are most appropriate. Generally, household testing programmes should be linked to source water testing programmes to ensure that the monitoring team understand whether poor quality in the home results from re-contamination or poor source water quality.

The numbers of samples and the selection of households will depend largely on the principal objective for the testing of household water. If the major purpose is to simply undertake random sampling of household water (which may be an important part of the monitoring programme whether "year group" or longitudinal study) then a stratified random sampling approach can be adopted. In this case, no specific intervention is being evaluated although the collection of information about sources and the type and cleanliness of the storage container may indicate where major problems lie.

This data can also be used to check on the use of feedback of surveillance results on household water quality. For instance in Uganda, the simple process of feedback of information and routine testing led to observable improvements in water quality stored within the home (Howard and Luyima, 2000). When such programmes are initiated it is important that different households are visited in each sampling period to prevent a bias developing due to repeated visits by surveillance staff. However, a cluster sampling approach may be adopted by identifying sentinel communities believed to be at greater risk because they have least access to direct connection or because they are more affected by interruption in supply.

In some cases there may be other specific objectives for testing water in the home. These may include evaluating the impact of a particular health education programme or household water storage and treatment interventions. In this case, a study would be designed to measure the impact between an intervention group and a control group thus allowing an evaluation of the impact of the intervention. Alternatively, the influence of the type of source, frequency and duration of discontinuity, or type of storage vessel on household water quality may be assessed in a community. In this case, a cluster sampling approach would typically be used to keep the number of households included to a reasonable number that allows intensive investigation.

## 6.6    Summary

This chapter provides a summary of possible sources of faecal contamination in drinking water and describes recognised sampling regimes in order to detect contamination. The small volume of samples collected during a reporting period represents only a tiny fraction of the total quantity of water delivered during that period. Thus, the challenge of sampling is to provide maximum information on water quality in the distribution system using the data from a limited number of samples. The ingress of faecal contamination into the distribution system should trigger immediate responses and this is covered in Chapter 7. The approaches available for sampling water quality from non-piped systems are also covered, with indications provided for actions to improve water quality.

# REFERENCES

Adriaanse, M. (1997) Tailor-made guidelines: a contradiction in terms? *European Water Pollution Control* **7**(4), 11-16.

Ahmed, F. and Hossain, M.D. (1997) The status of water supply and sanitation access in urban slums and fringes of Bangladesh. *Journal of Water Supply: Research and Technology – Aqua* **46**, 14-19.

Ainsworth, R.A. (2002) Water quality changes in piped distribution systems. World Health Organization.

Anon (1994) *The Microbiology of Water 1994: Part 1 – Drinking water.* Reports on Public Health and Medical Subjects, No. 71. Her Majesty's Stationery Office, London.

APHA, AWWA, WEF (1998) *Standard Methods for the Examination of Water and Wastewaters. 20th Edition.* American Public Health Association, Washington DC.

ARGOSS (2001) Guidelines for assessing the risk to groundwater from on-site sanitation. BGS Commissioned Report CR/01/142, British Geological Survey, Wallingford, UK.

AWWA (1986) *AWWA Standard for Disinfecting Water Mains.* AWWA C651-86, Denver CO.

Barrett, M.H., Johal, K. Howard, G., Pedley, S. and Nalubega, M. (2000) Sources of faecal contamination in shallow groundwater in Kampala. In: *Groundwater: Past Achievements and Future Challenges.* Sililo *et al.* (Eds.). pp 691-696.

Bartram, J. (1999) Effective monitoring of small drinking water supplies. In: *Providing Safe Drinking Water in Small Systems: Technology, Operations and Economics.* Cotruvo, J.A., Craun, G.F. and Hearne, N. (Eds.) Lewis Publishers, Washington, DC. USA, pp.353-366.

Bartram, J. and Ballance, R. (1996) *Water Quality Monitoring.* Chapman and Hall, London, UK.

Breslin, E. (2000) Protecting drinking water: water quality testing and PHAST in South Africa. In: *Water, Sanitation and Health*. Chorus, I., Ringelband, U., Schlag, G. and Schmoll, O. (Eds.) IWA, London, pp 89-94.

Burlingame, G.A. and O'Donnell, L. (1993) *Coliform sampling at routine and alternate taps: problems and solutions*. Proceedings of the AWWA Water Quality Technology Conference, Miami, FL.

Burlingame, G.A. and Choi, J.J. (1998) *Philadelphia's guidelines for obtaining representative samples from throughout drinking water systems*. Proceedings of the AWWA Water Quality Technology Conference, November 1-4, 1998, San Diego, CA.

Byappanahalli, M.N. and Fujioka, R.S. (1998) Evidence that tropical soil can support the growth of *Escherichia coli*. *Water Science and Technology* **38**(12), 171-174.

Clark, J.A. (1990) The Presence-Absence test for monitoring drinking water quality. In: *Drinking Water Microbiology*. McFeters, G.A. (Ed.) Springer-Verlag, New York, pp 399-411.

Clark, J.A. (1980) The influence of increasing numbers of non indicator organisms by the membrane filtration and P-A tests. *Canadian Journal of Microbiology* **15**, 827-832.

Craun, G.F. and Calderon, R.L. (2001) Waterborne disease outbreaks caused by distribution system deficiencies. *Journal of the American Water Works Association* **93**, 64-75.

Dufresne, L., Burlingame, G., Cochrane, C., Maley, L. Shahid, S. and Toch, M. (1997) *Eliminating 'noise' in distribution system coliform monitoring*. Proceedings of the AWWA Water Quality Technology Conference, Denver, CO.

Forget, G. and Sanchez-Bain, W.A. (1999) Managing the ecosystem to improve human health: Integrated approaches to safe drinking water. *International Journal of Occupational and Environmental Health* **5**(1), 38-50.

Gale, P. (1996a) Coliforms in the drinking water supply: What information do the 0/100-mL samples provide? *Journal of Water Supply: Research and Technology - Aqua* **45**(4), 155-161.

Gale, P. (1996b) Developments in microbiological risk assessment models for drinking water: A short review. *Journal of Applied Bacteriology* **81**, 403-410.

Gardner, J.F. and Peel, M.M. (1986) *Introduction to Sterilization and Disinfection*. Churchill Livingstone, Melbourne.

Geldreich, E.E. (1996) *Microbial Quality in Water Supply Distribution Systems*. CRC Press Inc. Boca Raton, FL.

Geldreich, E.E. and LeChevallier, M. (1999) Microbiological quality control in distribution systems. In: *Water Quality and Treatment: A Handbook of Community Water Supplies. Fifth Edition*. Letterman, R.D. (Ed.) McGraw-Hill, New York. pp.18.1-18.49.

Gelinas, Y., Randall, H., Robidoux, L. and Schmit, J-P. (1996) Well water survey in two Districts of Conakry (republic of Guinea) and comparison with the piped city water. *Water Resources* **30**(9), 2017-2026.

Haas, C.N. and Heller, B. (1986) Statistics of enumerating total coliforms in water samples by membrane filter procedures. *Water Research* **20**(4), 525-530.

HMSO (1989) The Water Supply (Water Quality) Regulation 1989. Statutory Instrument No. 1147.

Howard, G., Bartram, J.K. and Luyima, P.G. (1999) Small water supplies in urban areas of developing countries. In: *Providing Safe Drinking Water in Small Systems:Technology, Operations and Economics*. Cotruvo, J.A., Craun, G.F. and Hearne, N. (Eds.) Lewis Publishers, Washington, DC. USA. pp.83-93.

Howard, G. and Luyima, P.G. (2000) Report on water supply surveillance activities in 10 selected urban areas of Uganda. Report published for the Ministry of Health, Uganda and available at www.lboro.ac./watermark

Howard, G., Bartram, J., Schuab, S., Deere, D. and Waite M. (2001a) Regulation of microbiological quality in water cycle. In: *Water Quality: Standards, Guidelines and Health. Assessment of risk and risk management for water-related infectious disease*. Fewtrell, L. and Bartram, J. (Eds.) IWA Publishing, London, pp. 377-393.

Howard, G., Barrett, M., Pedley, S., Johal, K. and Nalubega, M. (2001b) Distinguishing human and animal faecal contamination in shallow groundwater. Unpublished technical report for DFID, Robens, Guildford.

ISO (1991) Sampling – Part 5: Guidance on sampling drinking water and water used for food and beverage preparation. ISO 5667-5. International Organization for Standardization, Geneva, Switzerland.

Kerneis, A., Nakache, F., Deguin, A. and Feinberg, M. (1995) The effects of water residence time on the biological quality in a distribution network. *Water Research* **29**(7), 1719-1727.

LeChevallier, M. (1999) The case for maintaining a disinfectant residual. *Journal of the American Water Works Association* **91**(1), 86-94.

Lloyd, B. and Bartram, J. (1991) Surveillance solutions to microbiological problems in water quality control in developing countries. *Water Science and Technology* **24**(2), 61-75.

Lloyd, B., Bartram, J., Rojas, R., Pardon, M., Wheeler, D. and Wedgewood, K. (1991) *Surveillance and improvement of Peruvian drinking water supplies.* University of Surrey, Guildford, UK.

MacKenzie, W.R., Hoxie, N.J., Proctor, M.E., Gradus, M.S., Blair, K.A., Peterson, D.E., Kazmierczak, J.J., Addis, D.G., Fox, K.R., Rose, J.B. and Davis, J.P. (1994) Massive waterborne outbreak of *Cryptosporidium* infection associated with a filtered public water supply, Milwaukee, Wisconsin, March and April 1993. *New England Journal of Medicine* **331**(3), 161-167.

Martin, R.S., Gates, W.H., Tobin, R.S. *et al.* (1982) Factors affecting coliform bacteria growth in distribution systems. *Journal of the American Water Works Association* **74**, 34-37.

Melian, R., Myrlian, N., Gouriev, A., Moraru, C. and Radstake, F. (1999) Groundwater quality and rural drinking water supplies in the Republic of Moldova. *Hydrogeology Journal* **7**, 188-196.

Ongley, E. (1998) Modernisation of water quality programmes in developing countries: issues of relevancy and cost efficiency. *Water Quality International* September/October, 37-42.

Pedley, S. and Howard, G. (1997) The public health implication of groundwater microbiology. *Quarterly Journal of Engineering Geology* **30**(2), 179-188.

Pipes, W.O., Ward, P. and Ahn, S.H. (1977) Frequency distributions for coliform bacteria in water. *Journal of the American Water Works Association* **69**, 664-668.

Pipes, W.O. and Christian, R.R. (1984) Estimating mean coliform densities of water distribution systems. *Journal of the American Water Works Association* **76**, 60-64.

Prévost, M., Rompré, A., Baribeau, H., Callier, J. and Lafrance, P. (1997) Service lines: their effect on microbiological quality. *Journal of the American Water Works Association* **89**(7), 78-91.

Quick, R.E., Venczel, L.V., Mintz, E.D., Soleto, L., Aparicio, J., Gironaz, M., Hutwagner, L., Greene, K., Bopp, C., Maloney, K., Chavez, D., Sobsey, M. and Tauxe, R.V. (1999) Diarrhoea prevention in Bolivia through point-of-use water treatment and safe storage: a promising new strategy. *Epidemiology and Infection* **122**, 83-90.

Rahman, A. (1996) Groundwater as a source of contamination in rapidly growing megacities of Asia: Case of Karachi, Pakistan. *Water Science and Technology* **34**(7-8), 385-292.

Reiff, F.M., Roses, M., Venczel, L., Quick, R. and Witt, V.M. (1996) Low-cost safe water for the world: A practical interim solution. *Journal of Public Health* **17**(4), 389-408.

Rojas, R., Howard, G. and Bartram, J. (1995) Groundwater quality and water supply in Lima, Peru. In: *Groundwater Quality*. Nash, H. and McCall, G.J.H. (Eds.) AGID Report No. 16. Chapman and Hall, London. pp. 159-167.

Semenza, J.C., Roberts, L., Henderson, A., Bogan, J. and Rubin, C.H. (1998) Water distribution system and diarrheal disease transmission: a case study in Uzbekistan. *American Journal of Tropical Medicine and Hygiene* **59**(6), 941-946.

Sobsey, M.D. (2002) Household water treatment and storage as appropriate technology for the developing world. World Health Organization.

UK Water Industry Research (1995) Microbiological growth in domestic pipework and fittings: Cambridge Workshop. Report DW 04/BUKWIR, 1995.

VanDerslice, J. and Briscoe, J. (1993) All coliforms are not created equal: a comparison of the effects of water source and in-house water contamination on infantile diarrheal disease. *Water Resources Research* **29**(7), 1983-1995.

Whittington, D., Lauria, D.T. and Mu, X. (1991) A study of water vending and willingness to pay for water in Onitsha, Nigeria. *World Development* **19**,179-198.

WHO (1996) *Guidelines for Drinking water Quality, Volume 2: Health and supporting criteria. Second Edition.* World Health Organization, Geneva.

WHO (1997) *Guidelines for Drinking water Quality. Volume 3: Surveillance and control of community supplies. Second Edition.* World Health Organization, Geneva.

WHO and UNICEF (2000) *Global Water Supply and Sanitation Assessment 2000 Report.* WHO/UNICEF, USA.

Wright, R.C. (1986) The seasonality of bacterial quality of water in a tropical developing country (Sierra Leone). *Journal of Hygiene* **96**(1), 75-82.

*Chapter 7*

# SURVEILLANCE AND INVESTIGATION OF CONTAMINATION INCIDENTS AND WATERBORNE OUTBREAKS

*P.R. Hunter, Y. Andersson, C.H. Von Bonsdorff, R.M. Chalmers, E. Cifuentes, D. Deere, T. Endo, M. Kadar, T. Krogh, L. Newport, A. Prescott and W. Robertson*

## 7.1    Introduction

This chapter examines the investigation of possible waterborne outbreaks (due to drinking water) and, in particular, the role of laboratory analyses in the investigation. Outbreaks are the most obvious manifestation of waterborne disease, though not all such disease is associated with outbreaks. The detection and investigation of outbreaks provides some of the best insights into the microbial aetiology and the types of process failures that lead to waterborne disease. As such, they provide essential information for hazard analysis and risk assessment associated with drinking water (see Chapter 3). Because of this, it is essential that outbreaks are adequately investigated so that the appropriate lessons can be learned and preventative measures applied to mitigate against future outbreaks and to improve the microbial safety of water generally.

The World Health Organization's (WHO) definition of a food - or waterborne - outbreak is when two or more persons experience a similar illness after ingestion of the same type of food or water from the same source and when the epidemiological evidence implicates the food or the water as the source of the illness (Schmidt, 1995). Unfortunately, in the early stages of an outbreak it is usually far from clear whether cases are linked or related to drinking water. This is a particular problem for common infections transmitted through various different routes. A small number of cases associated with a water supply may not be detectable against the general background of infection.

Instead, existing surveillance systems only detect general changes in the incidence of infectious disease.

A more useful definition of a waterborne outbreak, for the purposes of active surveillance, is when more cases than would be expected are clustered, geographically and in time. In other words, are more cases being reported from a particular geographical location than would be considered normal? Clearly, in order to make this judgement, there has to be a system in place for the detection of cases of infection and an understanding of the expected frequency of reporting.

An outbreak needs to be in progress to be detected by a public health surveillance system. Preventing outbreaks occurring in the first place is the focus of the authorities responsible for supplying drinking water. A combination of a study of outbreaks combined with theoretical risk analysis can be used to predict scenarios that are likely to lead to water becoming unsafe. After examining waterborne outbreaks of illness in general, this chapter includes an overview of the role of indicator parameters in providing early warning of possible outbreak scenarios and the importance of having contingency plans in place to expedite corrective action. It then goes on to examine waterborne outbreak investigation in more detail.

**Table 7.1. Outbreaks of infectious illness linked to drinking water in the UK, 1991-2000**

(Adapted from Percival *et al.*, 2000)

| Water system | Disease | Number of outbreaks | Number of cases |
|---|---|---|---|
| Public supplies[1] | Cryptosporidiosis | 24 | >2955 |
| | Campylobacteriosis | 1 | 281 |
| | **Total** | **25** | **>3236** |
| Private supplies[2] | Gastroenteritis of unknown cause | 2 | 81 |
| | Campylobacteriosis | 7 | 162 |
| | Giardiasis | 1 | 31 |
| | Cryptosporidiosis | 3 | 74 |
| | Enterohaemorrhagic *E. coli* | 1 | 14 |
| | Mixed campylobacteriosis and cryptosporidiosis | 1 | 43 |
| | **Total** | **13** | **405** |

1. Public supplies are owned by commercial water utilities.
2. Private supplies are not owned by commercial water utilities and vary from supplies providing water to single dwellings up to some quite large supplies.

## 7.2 Waterborne outbreaks

We have very little idea how many outbreaks of waterborne disease there are in the world as few countries, even in Western Europe and North America, have surveillance systems in place that can reliably detect such outbreaks (WHO, 1999). Two countries that do have good quality disease surveillance systems are the United States of America and the United Kingdom, both of which produce regular reports of the number of detected outbreaks associated with water.

Table 7.1 shows the number of outbreaks reported in England and Wales for the years 1991 to 2000, while Table 7.2 shows outbreaks for the USA for 1991 to 1998. From these two tables it is clear that outbreaks of illness associated with drinking water are common even in affluent nations and can be a cause of substantial illness. Furthermore, it can be seen that a relatively small number of pathogens have been implicated in these outbreaks.

**Table 7.2. Outbreaks of infectious illness (1991-1998) linked to drinking water in the USA**

(Moore *et al.*, 1993; Kramer *et al.*, 1996a; Levy *et al.*, 1998; Barwick *et al.*, 2000)

| Disease | Water system[1] Community Outbreaks | Cases | Non-community Outbreaks | Cases | Independent Outbreaks | Cases |
|---|---|---|---|---|---|---|
| Acute gastroenteritis of unknown cause | 5 | 10 105 | 35 | 4 673 | 3 | 51 |
| Giardiasis | 10 | 1 986 | 3 | 128 | 2 | 12 |
| Cryptosporidiosis | 6 | 407 637 | 2 | 578 | 2 | 39 |
| Norwalk-like virus | 2 | 742 | | | | |
| Campylobacteriosis | 1 | 172 | 2 | 51 | | |
| *Salmonella typhimurium* | 1 | 625 | | | | |
| Non-O1 *Vibrio cholerae* | 1 | 11 | | | | |
| *E. coli* O157 | 1 | 157 | 3 | 39 | 1 | 3 |
| *Shigella sonnei* | 1 | 83 | 5 | 484 | 1 | 5 |
| *Shigella flexneri* | | | | | 1 | 33 |
| *Plesiomonas shigelloides* | | | 1 | 60 | | |
| Hepatitis A | | | 1 | 46 | 1 | 10 |
| Total | 28 | 421 518 | 52 | 6 059 | 11 | 153 |

1. Community and non-community water systems are public water supplies that serve ≥15 service connections or an average of ≥25 residents for ≥60 days/year. A community water system serves year-round residents of a community, subdivision or mobile home park. A non-community water system can be non-transient or transient. Non-transient systems serve ≥25 of the same persons for >6 months of the year (*e.g.* factories or schools), whereas transient systems do not (*e.g.* restaurants, highway rest stations or parks). Independent systems are small systems not owned or operated by a water utility serving <15 connections or <25 persons.

In addition to those cases of illness associated with outbreaks, there remain an uncertain number of sporadic cases. A sporadic case is a single case of infection that is not obviously linked to other cases. In most sporadic cases of disease, it is usually impossible to state with certainty where that individual acquired his/her infection. Indeed, for most potentially waterborne diseases it is difficult to estimate the proportion of such sporadic cases that are associated with drinking water. What evidence there is comes from case-control and other epidemiological studies and these are reviewed elsewhere (Hunter, 1997).

Outbreaks of disease from drinking water supplies often result from chance events (Deere *et al.*, 2001). Table 7.3 provides an illustration of the diversity of scenarios that can results in drinking water outbreaks. This has significant implications for the design and operation of drinking water supplies. The water suppliers need to have preventative and emergency response procedures in place to ensure safe water delivery in the event of a variety of circumstances.

**Table 7.3. Scenarios affecting drinking water implicated in disease outbreaks**

(from Deere *et al.*, 2001)

| Causal event(s) | Aetiology | Water type | Cases | Reference |
|---|---|---|---|---|
| **Pre abstraction and treatment** | | | | |
| Surface run off from contaminated catchment after heavy rain. Increased chlorine demand due to turbidity | *Campylobacter* | Chlorinated surface water | 3 000 | Vogt *et al.*, 1982 |
| Contaminated surface run off from melt water and heavy rain entering municipal wells | *Campylobacter* | Untreated ground water | 241 | Millson *et al.*, 1991 |
| Drought followed by heavy rain agricultural surface run off and poor coagulation and mixing | *Cryptosporidium* | Chlorinated and package filtered river water | 34 | Leland *et al.*, 1993 |
| Poor mixing and flocculation with filters started up without backwashing | *Cryptosporidium* | Surface water (CT) | 13 000 | Rose *et al.*, 1997 |
| Increase in turbidity, poor coagulation and backwash recycling | *Cryptosporidium* | Surface water (CT) | 403 000 | Rose *et al.*, 1997 |
| Catchment contaminated by higher than realised population, chlorine dosage too low | *Giardia* | Chlorinated surface water | 350 | Shaw *et al.*, 1977 |

**Table 7.3. Scenarios affecting drinking water implicated in disease outbreaks**
*(continued)*

| Causal event(s) | Aetiology | Water type | Cases | Reference |
|---|---|---|---|---|
| **Post abstraction and treatment** | | | | |
| Backflow of farm contaminated river water due to low mains pressure | *Campylobacter* | Sand filtered groundwater | 2 000 | Mentzing, 1981 |
| Agricultural run-off entering unsealed supply | *Cryptosporidium* | Surface water (CT) | 27 | Badenoch, 1990 |
| Deliberate contamination of water storage tank | *Giardia* | Municipal supply | 9 | Ramsay and Marsh, 1990 |
| Cross connection between pressure dropped potable and wastewater lines at pump wash | *Giardia* & *Entamoeba* | Surface water (CT) | 304 | Kramer *et al.*, 1996b |
| Sewage overflow entering pipes after repairs of ice breaks made without post chlorination | *E. coli* O157 | Municipal supply | 243 | Swerdlow *et al.*, 1992 |
| Birds entering water storage tank | *Salmonella* | Untreated ground water | 650 | Angulo *et al.*, 1997 |

CT: conventionally treated.

## 7.3    Preventing outbreaks

The variety of scenarios that can lead to outbreaks from drinking water has been illustrated in Table 7.3. Each water supply system is unique and, therefore, the scenarios that could lead to an outbreak can differ between supplies. The relevant authorities need to assess the risk of outbreaks from a range of scenarios for each specific supply, and then controls should be put in place to prevent such outbreaks occurring.

A 'water safety plan' can be developed to detail both the design of controls and the operating practices that would theoretically lead to the consistent provision of safe water (see Box 1.3). Such a plan would consider both nominal operating conditions and unusual events. A detailed discussion of a water safety plan is outside the scope of this chapter, however, an overview of how the responsible authorities would manage 'incidents' of suspected unsafe drinking water is given here.

### 7.3.1    *Incident management*

For the purposes of this section of the chapter the term 'incident' will be used to refer to any situation in which there is reason to suspect that water being supplied for drinking is, or is about to become, unsafe. Such a broad definition means that a variety of triggers can lead to an incident being declared.

Judicious use of indicator parameters can provide the earliest practical warning of the possibility that water may become unsafe. In other cases an incident might not be declared until health authorities detect an increase in disease and begin to question the safety of the drinking water supply. Incident triggers could include:

- Process indicators:
  - Inadequate performance of a sewage treatment plant discharging to source water.

  - Inadequate performance of drinking water treatment plant.

- Notification of chance events:
  - Spillage of a hazardous substance into source water.

  - Failure of power supply to a critical asset.

- Non-microbial indicator parameters:
  - Extreme rainfall in a catchment.

  - Detection of unusually high turbidity (source or treated water).

  - Unusual taste, odour or appearance of water.

- Microbial indicator parameters:
  - Measurement of unusually high faecal indicator densities (source or treated water).

  - Measurement of unusually high pathogen densities (source or treated water).

- Public health indicators:
  - Disease outbreak for which water is a suspect vector.

For the purposes of this chapter, two categories of incident will be discussed separately, namely:

- Specified incidents involving a pre-determined response to a nominated and routinely measured indicator trigger.

- Unspecified incidents involving a more general response, which is not fully pre-determined, to a range of possible triggers.


### 7.3.2    *Response to specified incidents*

Indicators of potentially unsafe water can be selected and systematically monitored throughout the water supply chain or cycle. Such indicators should yield information in good time to enable corrective action to prevent unsafe water being supplied. Alert levels can be set against which to compare observations. Alert levels would typically be just within critical limits of operation, outside of which confidence in water safety would be lost. Pre-determined corrective actions can be implemented once alert levels are exceeded. The corrective action (contingency) plans form part the specified aspects of the incident preparedness program.

Incident plans can have a range of alert levels. These can be minor, early warning, necessitating no more than additional investigation by a designated team, through to full emergency, requiring all available personnel and equipment. Major emergencies are likely to require the resources of organisations beyond the authority primarily responsible for supplying drinking water, particularly the health authorities.

Incident plans typically consist of items such as:

- Accountabilities and contact details for key personnel, often including several organisations and individuals.

- Lists of measurable indicators that might trigger incidents along with a scale of alert levels.

- Clear description of the actions required in response to alerts.

- Location and identity of the detailed standard operating procedures and required equipment.

- Location of backup equipment.

- Relevant logistical and technical information.

- Checklists, proformas and quick reference guides.

The plan may need to be followed at very short notice so standby rosters, effective communication systems and up to date training and documentation are required

*Case study: Incident response to turbidity indicator levels*

The incident preparedness program of the Sydney Catchment Authority includes detailed contingency plans for responding to indicators of poor source water quality. For example, the Authority monitors turbidity at many points in the source water supply system. It has developed an integrated bulk water supply system that provides a number of source water supply options. The turbidity of water entering critical reservoirs is monitored continuously. If there is an increase of > 5 NTU within three hours then an incident is declared and an alternative source water may be used. Filtered water turbidity is also monitored continuously and if it exceeds 1 NTU, an alternative source water will be selected.

### 7.3.3    *Response to unspecified incidents*

Some scenarios that lead to water being considered potentially unsafe might not be specifically identified within incident plans. This may be either because the events were unforeseen, or because they were considered too unlikely to justify preparing detailed corrective action plans. To allow for such events, a generalised water safety incident response plan can be developed. The plan would be used to provide general guidance on identifying and handling of incidents along with specific guidance on responses that would be applied to many different types of incident.

Rather than alert-level categories being pre-determined, a protocol for situation assessment and declaring incidents would be provided that includes personal accountabilities and categorical selection criteria. The selection criteria may include:

- Time to effect.

- Population affected.

- Nature of the suspected hazard.

Alert levels can vary, as they do for specified incidents, from minor through to full-scale emergencies. The preparation of clear procedures, accountabilities and equipment for the sampling and storing water in the event of an incident can be valuable for follow up epidemiological or other investigations, and the sampling and storage of water from early on during a suspected incident should be part of the response plan.

The success of unspecified incident responses depends on the experience, judgement and skill of the personnel operating and managing the drinking water supply systems. However, generic activities that are common to many suspected contamination events can be incorporated within general unspecified incident preparedness programs. For example, for piped systems, emergency flushing standard operating procedures can be prepared, and tested, for use in the event that contaminated water needs to be flushed from a piped system. Similarly, standard operating procedures for rapidly changing or by-passing reservoirs can be prepared, tested and incorporated. The development of such a 'toolkit' of supporting material limits the likelihood of error and speeds up responses during incidents.

*Case study: General incident response involving emergency flushing*

Sydney Water has developed a detailed incident preparedness program. If it is suspected that water may be contaminated for whatever reason, an incident is declared. Among the emergency standard operating procedures available for use during incidents are systematic emergency flushing plans. These have been developed to provide standard operating procedures for the most rapid practical removal of suspect water from the distribution system. The plans have been prepared in manageable sections in a ready-to-use format for supply direct to operations officers.

### 7.3.4    *Water avoidance and boil water orders*

In most water supply scenarios it is possible to:

- Terminate the supply of water.

- Advise (some or all) consumers to avoid consuming water.

- Advise (some or all) consumers to treat water, usually by boiling.

An incident preparedness program should include a thorough evaluation of the basis for calling such orders. The objective of the order should be taken in the public interest and typically involves a final decision by health authorities.

Even where drinking water contamination is suspected, the public interest is not always best served by making avoidance or disconnection orders. Research has shown that many people do not follow advice to boil their water, in part because of confusion over what to do (Angulo *et al.*, 1997; O'Donnell *et al.*, 2000: Willocks *et al.*, 2000). Furthermore, there is also evidence that boil water notices can have negative public health consequences through causing anxiety and also burns and scalds (Mayon-White and Frankenberg, 1989; Willocks *et al.*, 2000). If advice to boil water is issued then the incident team must be convinced of an ongoing risk to health of drinking tap water, which outweighs any risk from the boil water notice itself (Hunter, 2000a). Disability-adjusted life years can be used to provide a common currency to assist in this type of health-based decision-making (Murray, 1994). Financial considerations are also likely to be important. Turning off water supplies can have major economic consequences due to lost production and damaged equipment.

The relevant authorities should have a clear understanding of the accountabilities, circumstances and criteria regarding the calling of such orders. In addition, practical operating procedures should be in place. For example, procedures for rapid shutdown, or for alerting the public in the event of a water avoidance or boil order, should be thoroughly planned. Additionally, any incident management team intending to issue advice to boil water should be very clear at the outset about the criteria that will be used to lift the advice.

Emergency water supplies, such as the use of water tankers, can be maintained on standby at all times, or powers can be put in place to enable commandeering. Rapid notification procedures such as media, mail-drops and public address system vehicles need to be practical and available at any time. Systems that enable tracing of water from source to consumer can assist in better targeting of these types of responses to minimise the extent of their impacts.

## 7.4     Outbreak investigation

This section outlines the steps typically taken in the investigation of an outbreak of suspected waterborne illness in a developed country. The timely discovery of an outbreak and its cause normally involves a long series of events with different agencies involved. Consequently, the most effective outbreak investigations follow a sequence of activities, outlined below:

- **Planning**. Planning should address key issues about who should be involved in the investigation and what their roles should be. Outbreak plans should also address who will have lead responsibility for implementing the outbreak plan and who will have leadership of the management group.

- **Outbreak detection and confirmation**. Normally an increase in reports of illness or detection of particular pathogens in human samples is the first sign of an outbreak. Rarely, the first sign of a waterborne outbreak can also be a technical problem with the water source or in the water treatment or distribution. An important first step in any outbreak investigation is the confirmation of an apparent outbreak. Before an outbreak is officially declared, possible causes of error should be considered and excluded. Such causes of apparent outbreaks include laboratory false positives, the introduction of new laboratory methods and sudden changes in reporting behaviour (Casemore, 1992).

- **Outbreak description**. The first step in outbreak description is the derivation of the 'case definition'. A case definition is necessary to identify those cases that should and should not be included in subsequent analyses. The case definition should contain the presence of key symptoms and/or laboratory results, geographical location and date of onset or notification. There may be several case-definitions in use at a single time (*e.g.* one for a possible case and one for a confirmed case). In the beginning of the outbreak investigation a fairly wide definition is needed in order not to lose cases. Later during the investigation, when more information is revealed, the case definition can often be narrowed. When the case definition has been agreed, the next step is to identify how many people meet the case definition, by a process of 'case finding'. This may involve reviewing existing laboratory or other notification records, or involve proactive searching by contacting doctors, or possible cases themselves to identify cases that may not have been formally notified. It is important to find out when the outbreak started and identify the first case (primary or index case). The date at which each case fell ill (and sometimes even the hour) plotted as a graph (epidemic curve) gives valuable epidemiological information, and can provide a picture of the outbreak (*e.g.* a point source or a continuous outbreak). The geographical spread might give an idea about the cause of the outbreak. Cases can be plotted on a map to examine the possibility of clustering (*e.g.* households with the same community water or households situated on just one part the water distribution system or a private well). Age, sex and other socio-economic data may also give information about the likely causes.

- **Hypothesis generation**. Once sufficient information has been collected, a preliminary hypothesis as to the cause of the outbreak can be generated. Based on this, various remedial control measures may be suggested.

- **Hypothesis confirmation**. When the outbreak team has a hypothesis as to the cause of the outbreak, efforts are directed at proving or disproving this suggestion. There are three strands to this part of the investigation: further epidemiological investigations, further microbiological analyses of human and environmental samples and, in the case of a suspected waterborne outbreak, a sanitary inspection of the water treatment and distribution system. Epidemiological investigations at this stage will normally be of the case-control or cohort type. In these types of study, cases and controls (other individuals who were not ill) are interviewed and the responses analysed statistically to identify differences between the two groups (Hunter, 1997). The further microbiological analyses during outbreak investigation may include additional collection of human or environmental samples or more detailed characterisation of those samples. The sanitary inspections of the water treatment plant and distribution system are undertaken to collect evidence of failure in, or inadequate design of, the water treatment system. Such information is helpful in confirming the hypothesis of a water source for the outbreak. Evidence of what went wrong is also essential for informing the water supplier on how such failures and the consequent risk to public health can be prevented in future. The entire treatment and distribution system should be surveyed. Evidence of failure may be available in existing routinely collected data (Section 7.5) or become obvious only after enhanced monitoring or after surveying the treatment and distribution system.

- **Strength of association**. When all the evidence has been collected, the outbreak management team has to come to a conclusion about whether or not the suspect water supply was indeed the cause of the outbreak. Both the UK and USA have developed a form of scoring system that attempts to define the reliability of the conclusion of any association between water and disease. However, the two systems are not compatible as the UK classifies the strength of association between water and disease whilst the US system classifies the completeness of the investigation. There is a need for an internationally agreed system of classifying the strength of association between drinking water and disease in outbreaks. Both of these categorisations give considerable weight to analytical epidemiological studies, most commonly the case-control study, although recent evidence has suggested that these studies may be highly biased in those outbreaks where the possible cause has been made public. Such bias can lead to drinking water being falsely associated with an outbreak (Hunter, 2000b; Hunter and Syed, 2002).

## 7.5    Reviewing existing data

In any outbreak investigation where drinking water is suspected as the cause, one of the key sources of information are the records of the routine analyses of water quality, typically already held by the water supplier. Such a retrospective review of routine water quality data will seek evidence of reduction in source water quality, failure in water treatment and distribution and, rarely, evidence of the presence of the suspect pathogen in the treated water supply.

Routine bacteriological tests of drinking water are, in most countries, concentrated on parameters like *E. coli*, thermotolerant coliforms, total coliforms and heterotrophic plate-count bacteria which have simple analysis techniques. In some countries faecal streptococci (enterococci) and spores of sulphite-reducing clostridia are also included in the routine tests. Recently, some countries have started to include tests for *Cryptosporidium* in samples originating from, or influenced by, surface water but these involve rather expensive sampling procedures and analytical techniques, and only a few countries demand such tests to be done.

The most commonly available microbial results will normally be *E. coli* or thermotolerant coliforms. These species are used as an index of relatively fresh faecal contamination. In addition to the microbial tests, physicochemical water parameters such as turbidity, pH, chlorine residual, colour and organic matter may be monitored. Additionally, registration of failures in water treatment units, filters, dosing equipment, water pumps, distribution system, intake pipelines and so on, is of utmost importance for later investigation and determining the cause of the outbreak of illness.

Other useful parameters that may be monitored by the water supplier, or may be obtained from other sources include meteorological data (*e.g.* rainfall) and data on incidents that might affect water flow or water quality (*e.g.* floods, droughts, avalanches etc.).

Among data that are not usually monitored by suppliers on routine basis, but may be helpful if available are:

- Leakage from sewers or storm overflows affecting the water source.

- Traffic or industrial accidents with an effluent causing water pollution.

- Incidents that create low or negative pressure inside the drinking water pipelines (which may allow the ingress of polluted water).

### Case study: A pressure drop and cross connection

This case study relates to an outbreak of illness in Hungary in 1986. Over the course of the outbreak (which lasted two weeks) about 350 cases were detected, with 14 different pathogens detected from clinical samples (11 serotypes of *Salmonella*, 2 of *Shigella flexneri* and *E. coli* O124). When the outbreak first came to light, interviews revealed that all the cases had consumed drinking water at Szolnok railway station. Although no breakdown or failure in the drinking water system had been noted (other than a pressure drop that had affected the whole area), microscopic examination of water from the station showed the presence of large amounts of diatomaceous algae. The algae were identical to those previously detected in the river, indicating that untreated river water (containing the town's sewage) was present in the station drinking water. Later, bacteriological analysis of the same samples confirmed the microscopic examination, with 75% containing at least 80/100 ml thermotolerant coliforms. The fault causing the contamination was eventually shown to be a cross connection with an industrial water system using river water, with the valve connection probably being opened in response to the drop in pressure.

### 7.6    Enhanced monitoring including pathogen detection

Following the detection or suspicion of an outbreak, it may be appropriate to increase the amount of sampling over that normally undertaken for a particular supply. The reasons for this are two-fold: in order to provide further evidence that the water supply is the source of the outbreak and to identify the failure in treatment or distribution that led to the outbreak. Enhanced monitoring may involve:

- Taking more samples than normal from the same sites.

- Taking samples from elsewhere in the distribution system.

- Undertaking microbiological analyses that would not normally be undertaken (this may include monitoring for pathogens).

Increased sampling using standard methods at the routine sites may be useful to detect short-lived transient events. Small supplies may only be sampled infrequently, once a month or less. If such a small supply is implicated in an outbreak, sample frequency may be increased to one or more samples daily.

Increasing the number of sites where samples are taken may be useful for detecting localised problems within the distribution system and the greater number of samples may also improve the chances of detecting transient events. In this context, it may be appropriate to extend sampling to include:

- Livestock and potential sources of human pollution from within the catchment area.

- Source water, including wells that may not be currently used for extraction and sediment from storage reservoirs.

- Various critical points in the treatment plant, including backwash water from filter beds.

- Water and sediment from various points in the distribution system, including service reservoirs, pipelines and consumers taps.

- Stored water such as container water, ice, or filters if these are available.

One of the most powerful pieces of evidence implicating a water supply as the cause of an outbreak of infectious disease is the demonstration of the causative agent in the supply, especially in water pre-dating the event. Therefore, during most suspected waterborne outbreaks efforts will be made to isolate the pathogen from the water.

*Case study: Additional sampling and catchment investigations*

In the UK, an outbreak of cryptosporidiosis was linked to Thirlmere reservoir a surface water source that was chlorinated but not filtered prior to distribution (Hunter and Syed, 2001). An increase in the cases of illness in the areas served by Thirlmere, followed the detection of oocysts in a sample of treated water (34/10 litres). Oocysts isolated from the clinical samples were found to be type 2 *Cryptosporidium* (a zoonotic strain). A subsequent investigation revealed oocysts in sheep faeces within the reservoir catchment, which supported the hypothesis that the sheep were the ultimate source of the outbreak. However, the following year an outbreak affecting people resident in Glasgow was associated with another unfiltered surface water source. As in the Thirlmere outbreak, genotyping of clinical cases was type 2 and oocysts were detected in sheep faeces from around the catchment. However, when the sheep oocysts were typed they were found to be a novel genotype not previously found in man, suggesting that sheep may not have been the source (Chalmers *et al.*, 2002a). These two studies illustrate the value of linking molecular methods to the investigation of outbreaks.

*Case study: Analysis of stored water*

An outbreak of illness implicated a supply zone serving about 12 000 people. During the course of the outbreak 1 267 cases were identified (an attack rate of over 10%). Clinical sampling found a range of pathogens, but the dominant one was *Salmonella hadar*. Water sampling indicated heavy faecal contamination, although in most cases pathogens could not be detected, however, salmonellae were isolated from two water samples, one of which was from bottled tap water that had been stored in a patient's refrigerator. In both cases *S. hadar* (*i.e.* the strain implicated in the outbreak) was identified. The source of the outbreak was traced to ingress of sewage-contaminated groundwater through a poor weld on a new water main. Although the construction works on the new trunk main was suspected very early on and the ingress identified, isolation of *S. hadar* from both patients and water samples confirmed the waterborne nature of the outbreak.

### 7.6.1 *Pathogen detection*

There are several pathogens for which there are well proven methods available for detection in water in the international literature and in national or international standards (Anon, 1994). This is the case with several enteric bacteria, such as thermotolerant *Campylobacter spp.*, *Salmonella spp.* and *Vibrio cholerae.*

Although a fairly uncommon cause of waterborne outbreaks (Hunter, 1997), salmonellae seem to be quite easily isolated from suspect water sources during outbreaks. *Shigella spp.*, however, are a common cause of waterborne outbreaks world-wide but the detection of *Shigella* spp. from water using traditional methods is difficult because of the lack of methods of appropriate selectivity. The detection of pathogenic *E. coli* in implicated water is not usually attempted because of the difficulty in distinguishing them from non-pathogenic *E. coli*. The exception to this is during outbreaks of enterohaemorrhagic *E. coli*, which is a much more severe disease and has certain cultural characteristics to help distinguish *E. coli* O157 from other types.

Of the viruses shown to be present in faecally contaminated drinking water, the enterovirus group can be most easily detected but, in contrast with the name of this group, they do not generally cause enteric disease and are rarely involved in overt outbreaks. The exception being polioviruses, these were the cause of large outbreaks in the past when a waterborne mechanism was often supposed, although only one proven case is known (Farley *et al.*, 1984). Their role in causing low-level transmission through drinking water, however, is widely speculated upon. Although waterborne hepatitis A (HAV) outbreaks have frequently been reported, the detection of the virus in the water is generally not attempted because of the lack of available technique. Only since 1979 have techniques been developed for the propagation of HAV in cell culture and isolation from water samples (Provost and Hilleman, 1979). The only known example of successfully culturing HAV in parallel with unconventional methods from water that caused waterborne outbreak was described in the early 1990s (Divizia *et al.*, 1993).

*Case study: Isolation of Salmonella sp.*

Clinical samples from an outbreak of illness uniformly showed *S. typhimurium* (phage type 4, biotype 2) to be the causative agent. Food samples were negative and no common food source could be identified. The waterborne route was suggested by the exclusion of other possible routes and also some of the descriptive epidemiology. Water sampling during the outbreak was found to be acceptable in terms of coliform content and plate counts. As a result of the absence of faecal indicators the isolation of the outbreak strain of *Salmonella* in two water samples was initially dismissed as being due to faulty sampling technique. The public health authority took action and a boil water order and increased chlorination ended the outbreak. Final proof of the waterborne nature of the outbreak was not made until several months after it had ended. It was realised that shortly before the outbreak a family (served by a pit latrine rather than sewerage connection), living close to the pipeline connecting one of the supply wells (that provided unchlorinated water to the network) to a water tower had experienced illness caused by *S. typhimurium.* Investigations revealed that the pipeline close to the pit latrine had a crack in it, allowing the small-scale intrusion of contaminated groundwater. This example demonstrates the importance of not ignoring data and it also highlights the fact that pathogens may be present in disinfected drinking water in the absence of faecal indicators.

## 7.6.2    Molecular techniques

The chance of detecting a pathogen from an implicated drinking water source is often much improved using novel microbiological methods (especially molecular techniques – see Chapter 8 for more details), this is particularly true for viruses with no readily available or rapid cultural method. This group includes rotaviruses, astroviruses, caliciviruses, hepatitis A virus, Norwalk virus and other small round viruses (West, 1991). Traditional methods for the detection of viruses are based on tissue culture techniques that can take several weeks to perform. Direct polymerase chain reaction (PCR) methods although faster than conventional cell culture techniques are less sensitive than culture techniques with low levels of viral particles undetectable in environmental samples. Combined tissue culture and PCR methods offer major advantages over each individual method in that the detection of infectious virus is maximised and PCR inhibitors are removed. The assay greatly reduces the time needed to detect these organisms with times reduced to a few days. New developments in PCR technology may provide faster more sensitive detection and quantification of viral particles in the future.

Alternative methods for the identification/detection of potentially pathogenic bacteria include the use of *in situ* hybridisation and species-specific probes (Prescott and Fricker, 1999). This powerful technique enables organisms to be detected *in situ* within a few hours and can be adapted for use with any organism. With the advancement of micro-array and technologies several different probes targeting many different pathogens can be processed together. This could be invaluable for sample analysis during outbreak conditions.

Whether to undertake such a demanding examination has to be decided in each situation. In most cases, success is dependent on the ready availability of personnel with the relevant skills and resources. For many pathogens the best results are likely to be obtained by a national or regional reference laboratory specially practised in the detection of certain pathogens. It is most useful if an action plan for outbreak investigation is available, containing the necessary steps to be taken and laboratories to be contacted in case of emergency. In turn, reference laboratories should also have a contingency plan in order to deal with urgent requests to participate in waterborne outbreak investigations.

*Case study: Virus identification*

In Finland a waterborne outbreak of Norwalk-like virus in Heinävesi was attributed to an outbreak of illness (affecting 500 people) three months earlier in Kuopio, a town 70 kilometres upstream (Kukkula *et al.*, 1999). The sewage from Kuopio is discharged to a lake (from which Heinävesi takes its raw water), which at the time of the outbreak was iced over. Reverse transcriptase polymerase chain reaction analysis revealed virus in tap water samples in Heinävesi and also demonstrated that the virus was identical to those isolated from clinical samples in both outbreaks (Maunula *et al.*, 1999).

### 7.6.3    Negative results

Although pathogen detection is important in outbreak investigation, recovery of pathogens from drinking water is often unsuccessful even when a supply is strongly associated with an outbreak. Probably the most common cause of failure to detect an implicated pathogen is the time between contamination and subsequent infection and the time that the outbreak is detected and investigations commenced. A transient contamination event may lead to only temporary contamination of the supply. The chance of finding the pathogenic agent is also dependent on the method used, the organism's robustness in the water environment in general and its resistance to water disinfectants. Additionally, the ability to detect a pathogen in a water supply may be hampered by the common practice of performing a preventative super-disinfection, which is sometimes conducted prior to ensuring that appropriate sample(s) are taken for examination of the water, so destroying any remaining pathogens that may have been present.

Even if pathogenic agents are detected in the implicated drinking water, this may not always correlate with the clinical picture. In one outbreak, for example, both echo- and coxsackie viruses were isolated from water samples but the clinical picture implicated a different type of viral infection (Stenström, 1994), clearly where sewage contamination has occurred the detection of mixed pathogens is unsurprising. In this situation, isolation of a pathogen different to the one causing the outbreak could only be taken as evidence of inadequate water management.

### 7.6.4    Pathogen typing and strain characterisation

Even where it has been possible to detect a pathogen, in some cases it may be insufficient to identify the causative organism in human or environmental samples only down to the species level. Further characterisation may be vital in determining the source of contamination and a number of properties can be utilised, such as antibiotic resistance profiles. These can differentiate for example between human and non-human faecal sources as the bacteria infecting humans and livestock are often resistant to different antibiotics.

The traditional use of typing is to enable the investigators to determine whether strains isolated from different sources are indistinguishable or not. Another use is in the determination of virulence (*i.e.* if different strains within a species vary in their ability to cause illness). Finally, sometimes different strain types have different epidemiology as in the case of *Cryptosporidium parvum* where type 2 strains are zoonotic and type 1 are largely restricted to causing infection in humans.

When choosing any typing method there are a number of criteria that need to be considered (Hunter, 1991). These include:

- Typability (the proportion of strains that can be typed by that method).

- Reproducibility (the probability that if the same strain was re-tested it would give the same result).

- Discriminatory power (the ability of a method to distinguish between unrelated strains).

In addition, cost, ease of use and timeliness are important factors. There are many different typing methods described in the literature and the optimal method depends on the organism under investigation, the reasons for typing (whether as an aid to characterise a few strains associated with a hospital outbreak or an aid to surveillance within a country) and the resources available to the typing laboratory (both financial and technical expertise).

This section focuses principally on modern molecular typing methods (with more details in Chapter 8), which have been used increasingly since the early 1980s. However, typing methods have been used by microbiologists long before then. One of the most important 'traditional' techniques is serotyping, which is still the primary typing method used in the categorisation of a number of microorganisms including: *Salmonella* spp., *Shigella* spp., *E. coli*, and the enteroviruses (Threlfall and Frost, 1990; Hinton, 1985; Wenner, 1982). It works by discriminating between strains on the basis of their surface antigens. For bacterial pathogens the method usually involves mixing the strain under investigation with various sera and looking for agglutination. For viruses, the technique usually involves demonstrating loss of the ability to infect tissue culture cells after mixing with sera.

Other traditional typing methods include (Aber and Mackel, 1981):

- Bacteriophage typing where strains are discriminated according to their susceptibility to killing by bacteriophages. This method is still commonly used in typing of *Staphylococcus aureus* and various serotypes of *Salmonella* (Threlfall and Frost, 1990).

- Biotyping that discriminates on the basis of the requirements for selected nutrients to grow. Although widely used in the past for many different pathogens including *E. coli* (Hinton, 1985), biotyping has largely fallen out of favour. However, it may still have a role to play in laboratories with few resources.

- Resistotyping distinguishes between strains on the susceptibility to various antibiotics and other antimicrobial agents, usually known as antibiograms. Resistotyping based on antibiotic sensitivity patterns has a considerable advantage in that antimicrobial susceptibility testing is frequently undertaken to guide therapy and so the data is usually to hand. As a typing method, resistotyping comes into its own for the rapid identification of strains with unusual antibiograms.

- Bacteriocin typing is based on the production of, or susceptibility to, various bacteriocins (compounds produced by bacteria that inhibit the growth of other strains). This method was commonly used in the typing of *Pseudomonas aeruginosa* when it was known as pyocin typing (Pitt, 1988), it has now, however, been largely superseded by other methods.

A major problem with traditional methods is that they are frequently of low typability and discriminatory power. Furthermore, typing can often only be done within certain reference laboratories and sending strains away can lead to delay. Many of the modern molecular methods offer considerable advantages for typing and a number of DNA 'fingerprinting' techniques have been described, including the following:

- Restriction fragment length polymorphism (RFLP).

- Pulse field gel electrophoresis.

- Randomly amplified polymorphic DNA (RAPD).

These enable each isolate to be characterised by a unique set of banding patterns which can be used for species identification or for epidemiological purposes and are described in more detail in a case study below and in Chapter 8.

*Case study: Shigella typing*

In 1998, in Nagasaki Japan, there was a large outbreak of *Shigella sonnei* infection, with 470 confirmed cases and 821 epidemiological linked cases. The outbreak investigation started when six students were reported ill (five of whom were hospitalised); all of who had eaten lunch at the University cafeteria. Active case finding and a cohort study of University users (students, staff and visitors) was undertaken. This found that 25% of regular University users had symptoms that met the case definition.

Patient interviews provided no evidence of a common food, but consuming water on campus was suspected to be associated with illness. The campus was supplied from two shallow wells, with no water treatment other than chlorination. Disinfection, however, was thought to be inadequate with samples showing no evidence of residual chlorine. Additionally, microbial tests were positive for *Shigella sonnei*. The source of the contamination was traced to a leakage of raw sewage from a nearby sewerage pipeline (identified using a sodium chloride tracer). DNA fingerprinting, using pulse field gel electrophoresis revealed that the isolates of *Shigella sonnei* were identical from both clinical and water samples. The outbreak was halted by issuing instructions not to drink the campus water and then switching from the well source to a municipal supply.

*Case study: Cryptosporidium identification*

Recent advances in the application of molecular biological methods to *Cryptosporidium* have contributed much to knowledge of the epidemiology of cryptosporidiosis. Human disease is usually caused by *C. parvum*, in which two genotypes predominate. Genotype 1 is the anthroponotic genotype (type H) that is largely restricted to humans, and genotype 2 (type C) is the zoonotic genotype that causes both human and animal disease (Fayer *et al.*, 2000). Thus, the detection of genotype 1 is indicative of a human source of infection or contamination and genotype 2 of either an animal or a human source. Genotypes, and indeed some species, of *Cryptosporidium* cannot be differentiated microscopically. Characterisation of isolates using DNA amplification-based methods is advantageous over phenotypic methods since relatively few organisms are required (Gasser and O'Donoghue, 1999).

Molecular characterisation of *Cryptosporidium* has included analysis of repetitive DNA sequences, RAPD, direct PCR with DNA sequencing and PCR/RFLP analysis (Clark, 1999; Morgan *et al.*, 1999). The two distinct

*C. parvum* genotypes have been consistently differentiated at a variety of gene loci (Fayer *et al.*, 2000), including:

- *Cryptosporidium* Oocyst Wall Protein (COWP).

- Ribonuclease reductase.

- 18S rDNA (*syn.* small subunit ribosomal RNA).

- Internal transcribed rDNA spacers (ITS1 and ITS2).

- Acetyl-CoA synthetase.

- Dihydrofolate reductase-thymidylate synthase (dhfr-ts).

- Thrombospondin related adhesive proteins (TRAP-C1 and TRAP-C2).

- the $\alpha$ and $\beta$ beta tubulin.

- 70kDa heat shock protein (hsp70).

Application of genotyping techniques has also led to the characterisation of additional *Cryptosporidium* spp. and genotypes, and it has become clear that while the majority of human cryptosporidiosis is caused by *C. parvum*, other species are also found infecting both immunocompetent and immunocompromised patients (Fayer *et al.*, 2000; Chalmers *et al.*, 2002b). It is evident that some primer pairs are species specific, such as those for TRAP-C2 which is specific for *C. parvum* (Elwin *et al.*, 2001), while others cross react with related protozoan parasites, and that some PCR/RFLPs differentiate species/genotypes more readily than others (Sulaiman *et al.*, 1999). While PCR/RFLP is widely used for characterisation, and allows many specimens to be analysed and compared, only bases at the restriction enzyme sites are examined. Sequence analysis provides the 'gold standard' since all the bases within the target sequence at the locus are examined. The importance of sequence confirmation of RFLP patterns was illustrated by Chalmers *et al.* (2002) who identified a novel RFLP pattern, similar to *C. parvum* genotype 1, in the COWP gene of isolates from sheep, but sequence data clearly differentiated the isolate. Therefore, careful primer selection and PCR product analysis is required for detection and characterisation, particularly from environmental specimens where a wide range of cryptosporidia and other organisms may be present. It must also be noted that oocyst recovery and PCR methods from environmental samples, including water, have yet to be standardised.

The further differentiation of subtypes within *Cryptosporidium* genotypes provides additional resolution for epidemiological investigations (Glaberman *et al.*, 2001), and a variety of tracking tools are being investigated and evaluated. The discovery of a number of dinucleotide and trinucleotide repeats within the *Cryptosporidium* genome has enabled the application of microsatellite typing as a method for further segregation within the two divisions of *C. parvum* (Caccio, 2000). Blasdall *et al.* (2001) have exploited an apparent fortuitous juxtaposition of two non-coding genes within the genome of *Cryptosporidium*, yielding host-level resolution of *C. parvum* in a robust method where banding patterns appear stable over a number of years within a single herd. Sequence analysis of small double-stranded extra chromosomal RNAs in *C. parvum* (Xiao *et al.*, 2001) and of a highly polymorphic gene encoding a 60KDa glycoprotein (Strong *et al.*, 2000) also offer tracking tools, while analysis of single strand conformation polymorphisms is also being investigated as a subtyping tool (Gasser *et al.*, 2001).

The methods discussed above are already having a significant impact on the investigation of outbreaks of waterborne cryptosporidiosis. The ability to distinguish between the anthroponotic and zoonotic genotypes is a significant pointer towards identifying the possible source of pollution. Patel *et al.* (1998) were able to demonstrate that two outbreaks originally thought to be due to agricultural pollution were actually due to human sewage. However, typing based only on genotyping has a low discriminatory power. In the case of an outbreak with a zoonotic strain, genotyping alone will not enable investigators to determine which herds are most likely to be responsible for the contamination events. The methods for increased strain discrimination discussed above have the potential to answer these types of question. Improved strain discrimination can also improve the epidemiological investigation of an outbreak by improving case definition. Possible cases who are infected with a strain other than the outbreak strain could be excluded from analysis, and so reduce potential for bias.

## 7.7    Summary

Waterborne outbreaks are the most obvious manifestation of waterborne disease. Microbiological examinations have several roles in the investigation of such outbreaks. The finding of the causative pathogen in the water supply is among the best evidence of a link between a water supply and an outbreak of disease. However, for a number of reasons it is frequently impossible to obtain this piece of evidence. Novel molecular methods may offer a better chance of identifying pathogens in the water supply than traditional cultural methods. However, sensitivity remains low and even if sensitivity were increased no test will detect an organism that was flushed from the distribution system a week or more previously. Incident preparedness plans would ideally include provision for the appropriate collection and storage of samples of water during suspected waterborne disease outbreaks to assist with follow-up. However, investigation methods following an outbreak do not require the pathogen to be detected in the supply. The second major use of microbiological investigations is in demonstrating failure in optimal water treatment and distribution. Laboratory tests or process indicators assessed as part of the routine management of the supply may provide useful information, as may the results of increased sampling using standard coliform and thermotolerant coliform counts. Above all, microbiology is essential in the diagnosis of individual cases of infection in the human population. The human population remains the best monitor of certain threats to the water supply (*e.g.* for cryptosporidiosis). Increasingly, novel technologies are being used to type strains isolated from humans to confirm that cases are indeed part of an outbreak of infection. Such typing may also provide clues to the epidemiology of an outbreak as when strains from humans can be shown to be the same as strains isolated from the environment.

Finally, microbial and other indicator parameters provide valuable tools for alerting the responsible authorities to the possibility of water becoming unsafe. Judicious use of indicator parameters within the context of a systematic water safety plan should provide early warning of potential public health incidents, enabling a planned corrective response. A more generalised state of incident preparedness should reduce the public health risk even in the event of a scenario for which a specific corrective response has not been prepared.

# REFERENCES

Aber, R.C. and Mackel, D.C. (1981) Epidemiologic typing of nosocomial microorganisms. *American Journal of Medicine* **70**, 899-905.

Andersson, Y. and Bohan, P. (2001) Disease surveillance and waterborne outbreaks. In: *Water Quality: Guidelines, Standards and Health. Assessment of risk and risk management for water-related infectious disease.* Fewtrell, L. and Bartram, J. (Eds.) IWA Publishing, London. pp.115-133.

Angulo, F.J., Tippen, S., Sharp, D.J., Payne, B.J., Collier, C., Hill, J.E., Barrett, T.J., Clark, R.M., Geldreich, E.E., Donnell, H.D. and Swerdlow, D.L. (1997) A community waterborne outbreak of salmonellosis and the effectiveness of a boil water order. *American Journal of Public Health* **87**(4), 580-584.

Anon (1994) *The Microbiology of Water 1994: Part 1 – Drinking water.* Reports on Public Health and Medical Subjects, No. 71. Her Majesty's Stationery Office, London.

Badenoch, J. (1990) *Cryptosporidium* in water supplies. Report of the Group of Experts. Department of the Environment, Department of Health. Her Majesty's Stationery Office, London.

Barwick, R.S., Levy, D.A., Craun, G.F., Beach, M.J. and Calderon R.L. (2000) Surveillance for waterborne-disease outbreaks--United States, 1997-1998. *Morbidity and Mortality Weekly Report. CDC Surveillance Summaries* **49** (SS-4), 1-21.

Blasdall, S.A., Ongerth, J.E. and Ashbolt, N.J. (2001) Differentiation of *Cryptosporidium parvum* subtypes by a novel microsatellite-telomere PCR with PAGE. *Proceedings of Cryptosporidium from Molecules to Disease, 7-12 October 2001.* Esplanade Hotel Fremantle, Western Australia. Murdoch University, Perth.

Caccio, S., Homan, W., Camilli, R., Traldi, G., Kortbeek, T. and Pozio, E. (2000) A microsatellite marker reveals population heterogeneity within human and animal genotypes of *Cryptosporidium parvum*. *Parasitology* **120**, 237-244.

Casemore, D.P. (1992) A pseudo-outbreak of cryptosporidiosis. *Communicable Disease Report. CDR Review* **2**(6), R66-R67.

Chalmers, R.M., Elwin, K., Reilly, W.J., Irvine, H., Thomas, A.L. and Hunter, P.R. (2002a) *Cryptosporidium* in farmed animals: the detection of a novel isolate in sheep. *International Journal of Parasitology* **32**, 21-26.

Chalmers, R.M., Elwin, K. and Thomas, A. (2002b) Unusual types of cryptosporidia are not restricted to immunocompromised patients. *Journal of Infectious Diseases* **185**, 270-271.

Clark, D.P. (1999) New insights into human cryptosporidiosis. *Clinical Microbiology Reviews* **12**, 554-63.

Deere, D., Stevens, M., Davison, A., Helm, G. and Dufour, A. (2001) Management Strategies. In: *Water Quality: Guidelines, Standards and Health. Assessment of risk and risk management for water-related infectious disease*. Fewtrell, L. and Bartram, J. (Eds.) IWA Publishing, London. pp. 257-288.

Divizia, M., Gnesivo, C., Bonapasta, R.A., Morace, G., Pisani, G. and Pana, A. (1993) Hepatitis A virus identification in an outbreak by enzymatic amplification. *European Journal of Epidemiology* **9**, 203-208.

Elwin, K., Chalmers, R.M., Roberts, R., Guy, E.C. and Casemore, D.P. (2001) The modification of a rapid method for the identification of gene-specific polymorphisms in *Cryptosporidium parvum*, and application to clinical and epidemiological investigations. *Applied and Environmental Microbiology* **67**, 5581-5584.

Farley, K.R.J., Rutherford, G., Litchfield, P., Hsu, S.T., Orenstein, W.A., Schonberger, L.B., Bart, K.J., Lui, K.J. and Lin, C.C. (1984) Outbreak of paralytic poliomyelitis, Taiwan. *Lancet* **ii**,1322-1324.

Fayer, R., Morgan, U. and Upton, S.J. (2000) Epidemiology of *Cryptosporidium*: Transmission, detection and identification. *International Journal for Parasitology* **30**, 1305-1322.

Gasser, R.B. and O'Donoghue, P. (1999) Isolation, propagation and characterization of *Cryptosporidium. International Journal for Parasitology* **29**, 1379-1413.

Gasser, R.B., Zhu, X.Q., Caccio, S., Chalmers, R., Widmer, G., Morgan, U., Thompson, R.C.A., Pozio, E. and Browning, G.F. (2001) Genotyping *Cryptosporidium parvum* by single-strand conformation polymorphism analysis of ribosomal and heat shock gene regions. *Electrophoresis* **22**, 433-437.

Glaberman, S., Moore, J., Lowery, C., Chalmers, R.M., Elwin K., Rooney, P., Millar, C., Dooley, J., Lal, A.A. and Xiao, L. (2001) Investigation of three drinking water associated outbreaks of cryptosporidiosis in Northern Ireland using genotyping and subgenotyping tools. American Society for Tropical Medicine and Hygiene meeting, Atlanta, GA.

Hinton, M. (1985) The sub-specific differentiation of *Escherichia coli* with particular reference to ecological studies in young animals including man. *Journal of Hygiene* **95**, 595-609.

Hunter, P.R. (1991) A critical review of typing methods for *Candida albicans* and their applications. *Critical Reviews in Microbiology* **17**, 417-34.

Hunter, P.R. (1997) *Water-borne Disease: Epidemiology and Ecology.* Wiley, Chichester.

Hunter, P.R. (2000a) Advice on the response to reports from public and environmental health to the detection of cryptosporidial oocysts in treated drinking water. *Communicable Disease and Public Health* **3**, 24-27.

Hunter, P.R. (2000b) Modelling the impact of prior immunity, case misclassification and bias on case-control studies in the investigation of outbreaks of cryptosporidiosis. *Epidemiology and Infection* **125**, 713-718.

Hunter, P.R. and Syed, Q. (2001) Community surveys of self-reported diarrhoea can dramatically overestimate the size of outbreaks of waterborne cryptosporidiosis. *Water Science and Technology* **43**, 27-30.

Hunter, P.R. and Syed, Q. (2002) Recall bias in a community survey of self-reported gastroenteritis undertaken during an outbreak of cryptosporidiosis strongly associated with drinking water after much press interest. *Epidemiology and Infection.* In press

Kramer, M.H., Herwaldt, B.L., Craun, G.F., Calderon, R.L. and Juranek, D.D. (1996a) Surveillance for waterborne-disease outbreaks - United States, 1993-1994. *Morbidity and Mortality Weekly Report. CDC Surveillance Summaries* **45**(SS-1), 1-33.

Kramer, M.H., Herwaldt, B.L., Craun, G.F., Calderon, R.L. and Juranek, D.D. (1996b) Waterborne disease – 1993 and 1994. *Journal of the American Water Works Association* **88**(3), 66-80.

Kukkula, M., Maunula, L., Silvennoinen, E. and v. Bonsdorff, C-H. (1999) Outbreak of viral gastroenteritis due to drinking water contaminated by Norwalk-like viruses. *Journal of Infectious Disease* **180**, 1771-1776.

Leland, D., Acanulty, J., Keene, W. and Stevens, G. (1993) A cryptosporidiosis outbreak in a filtered-water supply. *Journal of the American Water Works Association* **85**(6), 34-42.

Levy, D.A., Bens, M.S., Craun, G.F., Calderon, R.L. and Herwaldt, B.L. (1998) Surveillance for waterborne-disease outbreaks - United States, 1995-1996. *Morbidity and Mortality Weekly Report. CDC Surveillance Summaries* **47**(SS-5), 1-34.

Maunula, L., Piiparinen, H. and v. Bonsdorff C.-H. (1999) Confirmation of Norwalk-like virus amplicons after RT-PCR by microplate hybridization and direct sequencing. *Journal of Virological Methods* **83**, 125-134.

Mayon-White, R.T. and Frankenberg, R.A. (1989) Boil the water. *Lancet* **ii**, 216

Mentzing, L.O. (1981) Waterborne outbreaks of Campylobacter enteritis in central Sweden. *Lancet* **ii**, 352-354.

Millson, M., Bokhout, M., Carlson, J., Speilberg, L., Aldis, R., Borczyk, A.Z. and Lior, H. (1991) An outbreak of *Campylobacter jejuni* gastroenteritis linked to meltwater contamination of a municipal well. *Canadian Journal of Public Health* **82**, 27-31.

Moore, A.C., Herwaldt, B.L., Craun, G.F., Calderon, R.L., Highsmith, A.K. and Juranek, D.D. (1993) Surveillance for waterborne disease outbreaks - United States, 1991-1992. *Morbidity and Mortality Weekly Report. CDC Surveillance Summaries* **42**(SS-5), 1-22.

Morgan, U.M., Xiao, L., Fayer, R., Lal, A.A. and Thompson, R.C.A. (1999) Variation in *Cryptosporidium*: towards a taxonomic revision of the genus. *International Journal for Parasitology* **29**, 1733-1751.

Murray, C.J.L. (1994) Quantifying the burden of disease: the technical basis for disability-adjusted life years. *Bulletin of the World Health Organization* **72**(3), 429-445.

O'Donnell, M., Platt, C. and Aston, R. (2000) Effect of a boil water notice on behaviour in the management of a water contamination incident. *Communicable Disease and Public Health* **3**, 56-59.

Patel, S., Pedraza-Diaz, S., McLauchlin, J. and Casemore, D.P. (1998) Molecular characterisation of *Cryptosporidium parvum* from two large suspected waterborne outbreaks. *Communicable Disease and Public Health* **1**, 231-233.

Percival, S.L., Walker, J.T. and Hunter P.R. (2000) *Microbiological Aspects of Biofilms and Drinking Water*. CRC Press, Boca Raton.

Pitt, T.L. (1988) Epidemiological typing of *Pseudomonas aeruginosa*. *European Journal of Clinical Microbiology and Infectious Diseases* **7**, 238-247.

Prescott, A.M. and Fricker, C.R. (1999) Use of PNA oligonucleotides for the *in situ* detection of *Escherichia coli* in water. *Molecular and Cellular Probes* **13**, 261-268.

Provost, P.J. and Hilleman, M.R. (1979) Propagation of human hepatitis A virus in cell culture in vitro. *Proc. Soc. Exp. Biol. Med.* **160**, 213-221.

Ramsay, C.N. and Marsh, J. (1990) Giardiasis due to deliberate contamination of water supply. *Lancet* **336**, 880-881.

Rollins, D.M. and Colwell, R.R. (1986) Viable but non-culturable stage of *Campylobacter jejuni* and its role in survival in the natural aquatic environment. *Applied and Environmental Microbiology* **52**(3), 531-538.

Rose, J.B., Lisle, J.T. and LeChevallier, M. (1997) Waterborne cryptosporidiosis: Incidence, outbreaks and treatment strategies. In: *Cryptosporidium and Cryptosporidiosis*. Fayer, R. (Ed.) CRC Press, Boca Raton.

Schmidt, K. (1995) WHO surveillance programme for control of foodborne infections and intoxications in Europe. Sixth report, 1990-92. BgVV, Berlin, p. 14.

Shaw, P.K., Brodsky, R.E., Lyman, D.O., Wood, B.T., Hibler, C.P., Healy, G.R., Macleod, K.I., Stahl, W. and Schultz, M.G. (1977) A community-wide outbreak of giardiasis with evidence of transmission by a municipal water supply. *Annals of Internal Medicine* **87**, 426-432.

Stenström, T.A. (1994) A review of water-borne outbreaks of gastroenteritis in Scandinavia. In: *Water and Public Health*. Golding, A.M.B., Noah, N. and Stanwell-Smith, R. (Eds.) Smith-Gordon, London. pp.137-143.

Strong, W.B., Gut, J. and Nelson, R.G. (2000) Cloning and sequence analysis of a highly polymorphic *Cryptosporidium parvum* gene encoding a 60-kilodalton glycoprotein and characterization of its 15- and 45- kilodalton zoite surface antigen products. *Infection and Immunity* **68**, 4117-4134.

Sulaiman, IM, Xiao, L., and Lal, A.A. (1999) Evaluation of *Cryptosporidium parvum* genotyping techniques. *Applied and Environmental Microbiology* **65**, 4431-4435.

Swerdlow, D.L., Mintz, E.D., Rodriguez *et al.* (1992) Severe life-threatening cholera in Peru: predisposition of persons with blood group O. Abstract 941. Program, 32$^{nd}$ Interscience Antimicrobial Agent Chemotherapy Conference, 267.

Threlfall, E.J. and Frost, J.A. (1990) The identification, typing and fingerprinting of *Salmonella*: laboratory aspects and epidemiological applications. *Journal of Applied Bacteriology* **68**, 5-16.

Vogt, R.L., Sours, H.E., Barrett, T., Feldman, R.A., Dickinson, R.J. and Witherell, L. (1982) *Campylobacter* enteritis associated with contaminated water. *Annals of Internal Medicine* **96**, 292-296.

Wenner, H.A. (1982) The enteroviruses: recent advances. *Yale Journal of Biology & Medicine* **55**, 277-82.

West, P.A. (1991) Human pathogenic viruses and parasites: emerging pathogens in the water cycle. *Society for Applied Bacteriology Symposium Series* **20**, 107S-114S.

Willocks, L.J., Sufi, F., Wall, R., Seng, C. and Swan, A.V. (2000) Compliance with advice to boil drinking water during an outbreak of cryptosporidiosis. *Communicable Disease and Public Health* **3**, 137-138.

WHO (1999) *Water and Health in Europe*. World Health Organization Regional Office for Europe, Copenhagen.

Xiao, L., Limor, J., Bern, C. and Lal, A.A. (2001) Tracking *Cryptosporidium parvum* by sequence analysis of small double-stranded RNA. *Emerging Infectious Diseases* **7**, 141-145.

*Chapter 8*

# ANALYTICAL METHODS FOR MICROBIOLOGICAL
# WATER QUALITY TESTING

*W. Köster, T. Egli, N. Ashbolt, K. Botzenhart, N. Burlion, T. Endo, P. Grimont,
E. Guillot, C. Mabilat, L. Newport, M. Niemi, P. Payment, A. Prescott,
P. Renaud and A. Rust*

## 8.1    Introduction

There is a wide range of microorganisms of interest in water quality testing. Here we describe the general suite of methods currently used for the major indicator organisms and many of the pathogens of concern.

A fundamental limiting factor in the assessment of microbial quality of waters, and especially drinking water, is often the very low number of each organism present. Therefore, it is important to note that most microbiological procedure consists of: concentration/enrichment, detection and quantification (Table 8.1). A consequence of this multi-step approach is that technological advancement of any one step (such as detection) while possibly revolutionary, may be of limited value if the target group can not be satisfactorily concentrated before being subjected to the detection system.

This chapter is organised around the logical sequence of these method steps and common approaches for different microbial groups are discussed in one section. Emerging technologies are also presented, including the possible automation of the complete method or part of it. Performance and validation of methods and the statistical considerations behind choosing sample numbers are examined. The chapter concludes with a summary tabulation of the major methods along with their advantages and disadvantages and a list of abbreviations.

**Table 8.1. Example of the various method steps involved in the analysis of microorganisms**

| Common method components | Microbial groups | | |
|---|---|---|---|
| | Viruses | Bacteria | Parasitic protozoa |
| Concentration | Adsorption-elution | Membrane filtration | Cartridge filtration/IMS separation |
| Detection/ enumeration | Cell culture/ cytopathic effect, count plaque forming units | Selective growth on agar, count colony forming units | Immunological staining/count fluorescent cysts |

IMS: Immunomagnetic separation.


## 8.2 Recovery of target microorganisms

Traditional approaches to the isolation of microbial indicators have relied on various agar plate and liquid media methods. The basic pour plate technique has a maximum sample volume of about 1 ml whereas the spread plate technique uses 0.1 or 0.2 ml samples. For larger volume processing and rapid throughput, however, the membrane filtration technique is preferred if interfering particles are not concentrated simultaneously. Liquid cultivation techniques, either for the detection of the target organism (presence/absence test) or quantitatively, using multiple tube techniques and most probable number (MPN) calculations, allow flexible sample volume range and the handling of turbid samples. In liquid cultivation techniques, small volumes of sample dilutions or up to ten litre samples can be used. The detection of target microorganisms by non-cultivation methods is also presented for enteric viruses and parasitic protozoa.


### 8.2.1 Filtration methods

Bacteria are generally recovered on 47 mm diameter membrane filters with porosities of 0.22 to 0.45 µm. Membrane filters may be incubated on solid media, pads soaked in liquid media or as a MPN system in enrichment broth.

Cysts of protozoan parasites can be recovered on similar membranes but with larger surfaces (up to 293 mm diameter) and porosities as high as 2 µm (Ongerth and Stibbs, 1987). For convenience, however, various cartridge filters are generally preferred to recover protozoan cysts from up to 100 l water samples even in the presence of some turbidity (USEPA, 1999). The co-concentration of non-target particulates can, in part, be removed by subsequent selective separation method(s) (such as immunomagnetic separation (IMS), gradient centrifugation or flow cytometry, outlined in sections 8.2.2, 8.2.4.2 and 8.2.5). In England and Wales treated water supplies (10 l samples) are, however, analysed using compressed foam filters. Such sampling and monitoring procedures have been specified in a number of documents published by the UK Drinking Water Inspectorate (DWI: http://www.dwi.gov.uk/regs/ crypto/index.htm). A French Standard has also come into force in 2001 (NF T90-455, Publication date: 2001-07-01: Water quality Detection and enumeration of Cryptosporidium oocysts and of Giardia cysts- Concentration and Enumeration method) and an ISO is currently in preparation (ISO CD 15553 Water Quality-Isolation and Identification of Cryptosporidium Oocysts and Giardia Cysts from Water).

### 8.2.1.1  *Virus adsorption-elution methods*

A number of techniques have been described for the recovery of viruses by approaches based on the filtration of test water through filter media to which the phages/viruses adsorb. The phages/viruses are afterwards released from the filter media into a small volume suitable for quantitative plaque assays or presence/absence testing. The principle involved is that viruses/phages carry a particular electrostatic charge that is predominantly negative at or near neutral pH levels. This charge can be modified to predominantly positive by reducing the pH level to about 3.5. At this pH level viruses/phages will adsorb to negatively charged filter media. The balance involved is rather delicate because the lower the pH the better the adsorption, but low pH levels inactivate phages/viruses, and the sensitivity of different phages and viruses to low pH levels differs. Hydrophobic interactions also seem to play a role in the adsorption process (APHA, AWWA, WEF 1998). After adsorption, a small volume of an organic solution at pH 9.5 or higher is passed through the filter to reverse the charge on the viruses/phages to negative. This results in the release of the viruses/phages and they can be detected by conventional methods.

Bacterial viruses can also be retained by membrane filters under acidic conditions in the presence of divalent or trivalent salts. Sobsey *et al.* (1990) developed a relatively simple, inexpensive and practical procedure for the recovery and detection of F-RNA coliphages using mixed cellulose nitrate and acetate membrane filters for analysis of 100 to 2 000 ml volumes of tap water and 100 to 300 ml volumes of surface water. The efficiency of recovery of seeded F-RNA phages from 100 ml samples of tap water was 49%, which gradually decreased with increasing test volume to 12% for 2 000 ml. The efficiency of recovery from 100 ml and 300 ml samples of surface water was 34% and 18%, respectively. Although the procedure has attractive features, it should be weighed up against direct plaque assays on 100 ml samples, and presence/absence tests on 500 ml samples, both of which have theoretical efficiencies of 100% (Grabow *et al.*, 1998). Test volumes of the latter assays can be increased without loss of efficiency, as will be discussed later. Nonetheless, negatively or positively charged cartridge filters of various compositions remain the preferred approach for the concentration of viruses (enteric or bacteriophages) from large volumes of water.

Alternatively, filter media which carry a positive charge and hydrophobic binding sites at neutral pH levels, may be used to sorb negatively-charged viruses/phages at neutral pH levels (Sobsey and Glass, 1980). A variety of membranes and filter systems is available, among the well known ones are CUNO 1-MDS Virosorb and CUNO Zeta Plus 50-S or 60-S electropositive filters and glass wool. Application of these and related positively-charged filters in procedures with a wide variety of modifications and variations have been used to recover enteric viruses and phages (Singh and Gerba, 1983; Goyal *et al.* 1987). Efficiencies in the recovery of the coliphages (MS2, ØX-174, T2 and T4) from 17 litre volumes of tap water, sewage and lake water ranged between 34 - 100 % with positively charged Zeta Plus filters, however, MS2 appeared to be poorly recovered (range 0.3-1.8 %) with glass wool (Grabow *et al.*, 1998).

Therefore, although poliovirus and related viruses are recovered to some degree under certain conditions, evidence has been presented that phage recovery may be poor, probably because of poor adsorption as well as inactivation by exposure to the pH extremes required for adsorption and/or elution (Seeley and Primrose, 1982; Grabow *et al.*, 1998).

*8.2.1.2    Ultrafiltration*

Ultrafiltration is based on the filtration of water through membranes of polysulphonate or related material with a nominal molecular weight cut-off limit of about 10 000 Daltons. Particles with a diameter of 0.02 µm or more fail to pass through these membranes. Hence, dissolved organic molecules pass through the pores of these membranes but viruses and phages are too large to do so. It is, therefore, a process in which viruses are physically retained. Filter systems include spiral wound and sheet membranes (against which the water is kept in motion by means of a recirculating pump) or stirring apparatus (to enhance the filtration rate and avoid clogging) and yield close to 100% recovery (Grabow *et al.*, 1993). Other commercially available systems consist of units in which filtration is enhanced by tangential flow through hollow fibres with a large total filtration surface area, with some as disposable modules (described for *Cryptosporidium* by Simmons *et al.*, 2001).

Advantages of ultrafiltration include high recovery efficiencies and viruses/phages are not exposed to pH extremes or other unfavourable conditions, which may affect their viability. Adsorption of viruses and phages to the membranes is minimal, and this can be reduced by pretreatment of the membranes with beef extract (Divizia *et al.*, 1989) or 1-2% Tween 80, which seems to block potential adsorption sites. The most important disadvantage is that the membranes clog rapidly which implies that the volumes of water that can be processes are restricted.

## 8.2.2    *Immunocapture*

Direct immunomagnetic separation (IMS) techniques involve incubation of magnetic beads that are coated with specific antibodies for a target organism (see Box 8.1), in a mixture of the cell suspension (*e.g.* a water sample). After incubation and efficient mixing of the particles with the sample, the target cells become bound to the magnetic beads. The particles are then separated from the rest of the suspension with the help of a magnetic particle separator and washed several times.

---

**Box 8.1. Immunology techniques**

A wide range of immunological methods, taking advantage of antibody-antigen interactions, is available, among them the enzyme immunoassays (EIA). EIA methods combine the specificity of antibody molecules with the amplification of antibody-antigen interactions by enzyme catalysis. Different EIA methods exist. Many assays are performed in the wells of microtitre plates in which the reactants are immobilised. Antigen in the sample may or may not be bound by a specific antibody immobilized on the surface (coating antibody). Direct assays employ specific antibody conjugated to enzyme (enzyme linked immunosorbent assay - ELISA), whereas in indirect assays (double antibody sandwich - DAS-ELISA) the antigen-specific detecting antibody is detected by an anti-immunoglobulin enzyme conjugate. A number of DAS-ELISA approaches take advantage of the strong interaction between biotin and avidin (or streptavidin). Biotinylated antibodies are easily detected by using a streptavidin-enzyme conjugate. The same conjugate may be used to detect a number of different antibodies.

---

Immunoaffinity methods in combination with antibody coated magnetic beads have been used to isolate a number of different organisms from water samples, including hepatitis A virus (HAV), group A rotaviruses, pseudomonads, *E. coli* O157:H7 and *Cryptosporidium parvum*. The isolation of bacteria from water can be improved using enrichment followed by IMS and plating on selective agar. Moreover, magnetic beads coated with antibodies that specifically recognise various surface exposed epitopes of a variety of target organisms are already commercially available.

A basic laboratory infrastructure would be an advantage but is not absolutely necessary. The assays are easy to perform in a few hours. In addition, purification "kits" based on the immunocapture principle exist for several organisms. Although the technique is simple and fast, the efficiency of the reaction relies on the specificity and affinity of the commercially available monoclonal antibody and on the turbidity of the water sample. Immunocapture-based methods can be used as sound basis for other detection techniques (such as polymerase chain reaction (PCR), reverse transcriptase-polymerase chain reaction (RT-PCR), flow cytometry and fluorescent in-situ hybridisation (FISH), covered in Sections 8.2.5, 8.3.2.1 and 8.3.2.2).

### 8.2.3    Flocculation

Relatively successful techniques are on record for the recovery of enteric viruses from water by adsorption of viruses to flocculants such as aluminium hydroxide (APHA, AWWA, WEF 1998). The process probably involves both electrostatic interactions between the negatively charged virus surface and the positively charged aluminium hydroxide surfaces and coordination of the virus surface by hydroxo-aluminium complexes. Flocs are generally recovered by centrifugation or filtration. The flocs are then disintegrated by vigorous shaking and the viruses recovered by centrifugation (APHA,AWWA,WEF, 1998). The procedure is suitable for the recovery of viruses from relatively small volumes (several litres of water). This has been confirmed in tests using ammonium sulphate supplemented with beef extract for flocculation which yielded efficiencies of recovery of up to 85% for phages MS2, ØX174 and T3 (Shields and Farrah, 1986). Modifications of the procedure include magnetic organic flocculation, in which casein flocs are formed in the presence of magnetite for subsequent collection of the flocs by means of a magnet. The recovery of coliphages from waste- and lake water by this procedure has been described (Kennedy et al., 1985).

A method for the concentration of particles in the *Cryptosporidium* oocyst size range from water has been developed based on the calcium carbonate flocculation (crystallisation) (Vesey et al., 1993). An aliquot volume of water sample is treated by adding solutions of calcium chloride and sodium bicarbonate and raising the pH value to ten with sodium hydroxide, resulting in the formation of crystals of calcium carbonate, which enmesh particles. The crystals are allowed to settle, the supernatant fluid is discarded and the calcium carbonate precipitate is dissolved in sulphamic acid. This process yields reproducibly high recovery rates. It has, however, been suggested that the oocysts may not be used for the viability test because solution of the calcium carbonate with sulphamic acid has been reported to affect viability measured by fluorescent dye exclusion (Campbell *et al.*, 1994).

## 8.2.4    Centrifugation

### 8.2.4.1    Continuous flow centrifuge

The most common separation method is that of differential centrifugation (pelleting) using either a swinging bucket or a fixed angle rotor. However, this conventional method is limited to small volumes of water. For harvesting microbes to be tested from source and drinking waters, continuous flow rotors are preferred as they allow efficient processing of large volumes of water in a single run regardless of turbidity of the sample water.

The basic instrument is a continuous flow rotor in combination with a refrigerated centrifuge and a simple peristaltic pump. Continuous flow experiments are normally carried out in the cold in order to avoid heating the particle concentrate. In practice, sample water is pumped in continuously through the centreline of the seal assembly of the rotor while it is spinning at operating speed. The sample flows along the bottom of the core and moves over the centripetal surface of a solution. The centrifugal separation therefore accounts for two fractions:

- A sedimenting particle that moves out into the rotor cavity.

- A supernatant fraction that continues to flow along the core and over the centripetal surface of the water, then out of the rotor via the outlet lines.

The sample particles are allowed to pellet on the rotor wall.

The continuous flow centrifuges currently commercially available are large and stationary, and are not suited to concentrating water samples on site. Recently, a compact, continuous flow centrifuge with disposable plastic bowls (a modified blood component separation system) has been applied to the concentration of *Cryptosporidium* oocysts and *Giardia* cysts from large volumes of water. The robustness and accuracy of this system has not yet been fully examined and further experiments are also needed to examine the reproducibility and ease of recovery of the microbes from the disposable plastic bowl.

### 8.2.4.2   Gradient density separation/isolation

A centrifugation technique is also commonly used for separation/isolation of microbes, such as *Cryptosporidium* oocysts and *Giardia* cysts, from particle concentrates. In this case, a density gradient within a medium is centrifuged, separating microbes/particles from a thick mixture based on their specific density. The density gradient method involves a supporting column of fluid (such as sucrose or Percoll) where the density increases either zonally or linearly toward the bottom of the tube. If the density gradient column encompasses the whole range of densities of the sample particles, each particle will settle only to the position in the centrifuge tube at which the gradient density is equal to its own density. Thus, resulting in the separation of particles into zones solely on the basis of their density differences, although with environmental samples, the density gradient centrifugation step may lead to more than 30% losses with oocysts or cysts.

It is sometimes easier to start with a uniform solution of the sample and the gradient material such as a self-generating caesium chloride gradient for virus purification. Under the influence of centrifugal force, the material redistributes in the tube so as to form the required density gradient. Meanwhile, sample particles, which are initially distributed throughout the tube, sediment or float to their isopycnic positions. The target microbes can be recovered by removing the required density zone from the centrifuge tube. Development of density markers, which can be mixed in a particle concentrate prior to centrifugation should easily differentiate the zone to be collected.

### Biohazard

Concentration or separation of pathogenic materials by preparative centrifugation is deemed a biohazard. Extreme precautions must be taken when such samples are used because of the possibility of seal leakage or rotor mishaps. There is no standard method for decontaminating rotors exposed to pathogenic materials. Rotors should be cleaned with appropriate detergents and/or disinfectants according to the manufacturer's instructions. The widely used method is autoclaving and most commercially available rotors can be autoclaved, although the instruction manuals should always be consulted to ascertain any specific handling requirements.

## 8.2.5    *Flow cytometry*

Flow cytometry is a technology in which a variety of measurements can be made on particles, cells, bacteria and other objects suspended in a liquid. In a flow cytometer, particles are made to flow one at a time through a light beam (laser beam) in a sensing region of a flow chamber. They are characterised by light scattering based on their size, shape and density and also on the dyes that are used either independently or bound to specific antibodies or oligonucleotides that endow a fluorescent phenotype onto components of interest. As a particle flows through the beam, both light scattered by the particle and fluorescence light from the labelled particle is collected either by a photomultiplier or photodiode in combination with light splitters (dicroic mirrors) and filters. This makes it possible to make multiple simultaneous measurements (up to six parameters) on a particle. A solid phase laser scanning analyser might be an alternative for the flow cytometry technology, though it is still in its infancy. In the latter system, the fluorescent dye-stained samples loaded on a membrane filter are scanned by a laser beam, and fluorescence emitted from the dye attached to the target particle is similarly measured.

For the concentration of target organisms, a flow cytometer with the additional capacity to selectively sort (such as fluorescently activated cell sorting [FACS]) any selected particle from the suspension can be used. The ability to sort particles is an important feature for environmental microbiology since it makes it possible to collect presumptive organisms and to confirm results by, for example, visual examination. However, incorporation of a sorting unit into the system not only doubles the cost of the basic instruments but is also problematic for the development of the automatic monitoring system. Alternatively, additional detection parameters, such as dual staining with a second monoclonal antibody can be used to determine that both antibodies are binding an authentic target organism. This results in an increase in sensitivity of the detection method to such a degree that nonsorting detection (analyser only mode) is possible, although this has yet to be applied in routine practice (Vesey *et al.*, 1994).

A particularly valuable aspect of flow cytometry is its capability of rapid analysis: the assay itself can be completed within three to five minutes. This is likely to be one of the key devices for the routine multiple monitoring of microbes of interest (including a variety of indicator or pathogenic microbes and even viable but non-culturable bacteria). Although the applicability of this system is very broad, the current application of flow cytometry for monitoring of drinking water is limited (Deere *et al.*, 2002).

The basic instrument is the flow cytometer, which requires a skilled operator. The main consumables (in cost terms) are primarily (monoclonal) antibodies. The first and foremost problem, affecting the use of flow cytometry technology in this field is the high capital cost.

The technological limitation of this system is the number of dye combinations that can be used, where the combinations of excitation and emission spectra must be significantly different. The number and variety of specific labelling reagents will be another limitation of the system. Availability of commercial kits is expected to increase the use of this technique in many research fields including safe drinking water. In addition, most of the pathogenic microbes to be measured occur in drinking water at very low concentrations. When a negative sample is analysed no particles should be detected and a sample seeded with an aliquot of organisms should have an exact number of particles added. However, at present, it is difficult to obtain this level of sensitivity. Often a negative sample will contain some particles due to non-specific binding of antibodies to some interfering particles found in water samples, no matter how specific the antibody is.

*Biohazard*

Handling of the particle concentrate to be measured and the effluent from the flow cytometer is deemed to be a biohazard. Effluents must be autoclaved before discarding. The cytometer can be decontaminated (disinfected) between samples and at the end of the run by running 10% sodium hypochlorite (bleach solution) for 30 seconds and detergent solution for two minutes followed by a distilled water flush.

### 8.2.6    Pre-enrichment and enrichment techniques

As outlined in Chapter 2, detection and enumeration of index and indicator parameters rather than the search for specific pathogenic bacteria is used in routine bacteriological analysis of water. Nevertheless, under special circumstances the search for pathogenic bacteria may be necessary, *e.g.* during an epidemic (see Chapter 7) or when evaluating new water resources (WHO, 1984). Typically the number of pathogenic microorganisms is low (Emde *et al.*, 1992) and their recovery is low because they are in a stressed conditions. Therefore, the chances of detecting pathogenic bacteria will be greater by using a pre-enrichment step prior to enrichment and selective plating. This allows environmentally stressed organisms to recover and grow before selective pressures are applied. Generally, pre-enrichment media contain no antibiotics or other selective agents and this allows the growth of most microorganisms in the sample. Subsequent inoculation into enrichment media selects for the pathogen of interest, which can be detected by plating onto solid selective media. It should also be pointed out that this limits the ability to later quantify the pathogens in the sample (Ericksen and Dufour, 1986). Table 8.2 is based on the principles described above. Note that cell culture enrichment of viruses and phages is also used prior to detection by plaque assay (Grabow *et al.*, 1998) or PCR (as cell culture-PCR).

**Table 8.2. Procedures for the pre-enrichment and enrichment of bacterial pathogens using liquid media**

| Organisms | Enrichment conditions | Reference |
|---|---|---|
| *Yersinia enterocolitica* | "Cold enrichment" at 15°C in peptone-yeast extract broth. | Schiemann (1990) |
| | Selective medium, alkaline bile-oxalate-sorbose broth, pH 7.6. | Schiemann (1990) |
| *Salmonella* spp. | Pre-enrichment in buffered peptone water, then enrichment in, *e.g.* selenite-containing broth. | WHO (1984) |
| *Shigella* spp. | Enrichment medium, *e.g.* alkaline nutrient broth pH 8.0. | WHO (1984) |
| Cholera and non-cholera Vibrios | Enrichment medium, alkaline peptone water, or taurocholate tellurite peptone water. | WHO (1984) |
| *Legionella* spp. | Selective medium, charcoal yeast extract base amended with selected antibiotics. | States *et al.* (1990) |

## 8.2.7    Other techniques

### 8.2.7.1    Hydro-extraction

This procedure is based on placing a water sample into a cellulose dialysis bag, which is exposed to hygroscopic material such as polyethylene glycol (PEG). The PEG extracts water and micro-solutes through the semipermeable membrane while viruses and other macrosolutes remain inside. The procedure is recommended as an option for the recovery of viruses from small volumes of water, not more than a few hundred millilitres (APHA, AWWA, WEF 1998). The method may also be suitable for phages and has been used to recover cyanophages from ponds (Padan *et al.*, 1967).

### 8.2.7.2 Solvent-extraction

Solvent extraction is often applied as the initial step in separating viruses from solids, prior to polyethylene glycol precipitation, chromatography and guanidinium isothiocyanate (GIT) extraction (Shieh *et al.*, 1997).

## 8.3 Detection, identification and quantification of microorganisms

This section describes the more "classical" methods, which depend largely on cultivation techniques, as well as molecular methods. A number of them, particularly most of the recent techniques require standardisation and validation. Nonetheless, the majority of the methods presented here have already proven to be useful in drinking water microbiology and/or medical diagnostics, or display great potential.

In the detection, identification and quantification of target organisms some approaches are solely based on a single technique whereas other strategies take advantage of a combination of different methods. For example, to identify *Escherichia coli* reliance can be placed on a one-day-cultivation on chromogenic media. Alternatively, in a much faster approach, short pre-cultivation on an artificial medium can be combined with labelling using fluorescent probes, microscopy, and laser scanning techniques (section 8.4.1).

In the following sub-sections, alternative approaches are offered for a number of target organisms. The traditional cultivation techniques are usually sensitive but the identification is often not as reliable as might be desired. Methods based on molecular biology tend to be sensitive and yield reliable identification, but cultivation techniques always show viable organisms whereas molecular methods often reveal dead or inactivated target organisms/nucleic acid. This is of relevance in disinfected waters and should be considered in the interpretation of results.

### 8.3.1 Cultivation techniques

#### 8.3.1.1 Cultivation of bacteria

It has long been recognised that culture media lead to only a very small fraction (0.01 – 1 %) of the viable bacteria present being detected (Watkins and Xiangrong, 1997). Since MacConkey's development of selective media for *E. coli* and coliforms at the beginning of the 20th century, various workers have shown these selective agents inhibit environmentally or oxidatively stressed coliforms (McFeters *et al.*, 1986). Specially developed media without selective detergent agents (*e.g.* the m-T7 medium of LeChevallier *et al.*, 1982) permit a significant improvement in the recovery of stressed target bacteria. In addition, peroxides and superoxides are generated through auto-oxidation and photochemical reactions during the process of preparing, sterilising and storage of selective media (Lee and Hartman, 1989). Stressed cells have reduced catalase activity (Calabrese and Bissonnette, 1990) and are subject to additional stress once placed on selective media. Coupled with this is the accumulation of toxic hydrogen peroxide generated by aerobic respiration. Media without harsh selective agents have, therefore, taken over from the traditional approach (Hurst *et al.*, 2001).

Each of the cultivation techniques has a particular detection range depending on the sample volume. Whereas the lower detection limit depends on the maximum sample volume that can be processed, the upper limit can be freely chosen by selection of the dilution of the sample assayed. The measurement uncertainty related to each cultivation technique and statistical considerations based on Poisson distribution of target organisms in the sample have been described in documents produced by the Technical Committee on Water Quality of the International Organization for Standardization (ISO/TC 147/SC 4/WG 12).

The presence/absence test is sometimes used to monitor high quality samples where the presence of the target organism is improbable. It yields no information on the contamination level if a positive result is observed. The sensitivity of this technique depends on the sample volume analysed and the precision on the number of samples analysed in parallel at each dilution step. When using enough replicates good precision can be achieved. Computer programs now available for the calculation of MPN, give freedom to optimise the design without the restrictions of fixed MPN tables (Gonzales, 1996). In the techniques based on colony counting, the precision increases with increasing total number of colonies counted from replicate plates and from different dilutions. High densities of colonies on plates can cause overlap error and the interference of non-target colonies also limits the number of colonies to be reliably counted from one plate. Therefore, the upper working limit for a plate in colony counting techniques depends on the method (selectivity and distinction of the target), the target organism (size of target colonies), and the sample (background growth). In all of the enumeration techniques, the cultivation conditions are selected to promote the multiplication of the target organisms while simultaneously inhibiting the growth of other organisms. The balance between sensitivity and selectivity is the reason for different methods or sample processing for drinking water and highly contaminated waters.

Table 8.3 summarises the advantages and disadvantages of the commonly used cultivation techniques.

**Table 8.3. Established cultivation techniques**

| Technique | Advantages | Disadvantages |
|---|---|---|
| Most probable number (MPN) using liquid media | • Flexible sample volume range<br>• Applicable to all kinds of samples<br>• Allows resuscitation and growth of injured organisms<br>• Usually easy interpretation of test results and no special skills required<br>• Minimal time and effort needed to start the test<br>• The precision and sensitivity can be chosen by selection of volumes analysed, number of dilution levels and number of replicate tubes<br>• Media often inexpensive | • In routine application, when few replicates are used, the precision is often low<br>• Confirmation steps involving new cultivations are usually needed, which increase costs and time<br>• When the selectivity of the medium is not adequate, the target organisms can be masked due to the growth of other microorganisms<br>• Sample may contain inhibitors affecting the growth of the target organisms<br>• For the isolation of pure cultures, further cultivation on solid media is necessary<br>• If big sample volumes are studied costs of media increase and large space for incubation is needed |
| Presence/ absence test using liquid media | • As above | • As above<br>• No information on level of concentration of target organisms |
| Pour plate | • Simple and inexpensive method | • The sample volume analysed routinely is a maximum of 1 ml<br>• Thermal shock, caused when melted agar is poured on the sample, inhibits sensitive organisms<br>• Scoring of typical colonies not easy |

**Table 8.3. Established cultivation techniques** *(continued)*

| Technique | Advantages | Disadvantages |
|---|---|---|
| Spread plate | • Strictly aerobic organism are favoured because colonies grow on the agar surface (unless anaerobic conditions are applied) <br><br> • Differentiation of the colonies is easier than from pour plates | • The sample volume analysed routinely is a maximum of 0.1 ml <br><br> • Scoring of typical colonies not always easy |
| Membrane filtration | • Flexible sample volume range enabling the use of large sample volume and therefore increased sensitivity <br><br> • Water soluble impurities interfering with the growth of target organisms separated from the sample in the filtration step <br><br> • Quantitative result and good precision if the number of colonies grown adequate <br><br> • Further cultivation steps not always needed, which lowers the costs and time needed for the analysis <br><br> • When confirmation is needed, isolation from well separated colonies on membrane is easy | • Quality of membranes varies <br><br> • Solid particles and chemicals adsorbed from sample to the membrane during filtration may interfere with the growth of the target organism <br><br> • Not applicable to turbid samples <br><br> • Scoring of typical colonies not always easy |
| Liquid enrichment + confirmation and/or isolation on solid media | • Liquid enrichment in favourable media and incubation temperature allows resuscitation of injured or stressed cells <br><br> • Streaking of a portion of enrichment culture on an agar medium allows isolation of separate colonies <br><br> • Differentiation and preliminary identification is possible on selective solid media <br><br> • Detection and identification of organisms occurring in low numbers possible (*e.g. Salmonella*) | • Many cultivation steps increase costs of media, labour, skills needed and duration of the test |

251

## Chromogenic media-based detection methods

Media without harsh selective agents, but specific enzyme substrates allow significant improvements in recoveries and identification of target bacteria. In the case of coliforms and *E. coli*, such so-called 'defined substrate methods' were introduced by Edberg *et al.* (1991). This has evolved into the Colilert® technique and has been shown to correlate very well with the traditional membrane filter and MPN methods when used to test freshwater (Fricker *et al.*, 1997; Eckner, 1998). A number of such enzyme-based methods, allowing quantification within 24 hours is now available, including:

- Enterolert®, manufactured by IDEXX.

- Colisure® manufactured by IDEXX.

- Colilert®, manufactured by IDEXX.

- m-ColiBlue®, manufactured by Hach.

- ColiComplete®, manufactured by BioControl.

- Chromocult®, manufactured by Merck.

- MicroSure®, manufactured by Gelman.

The Colilert® method is based upon the sample turning yellow, indicating coliforms with β-galactosidase activity on the substrate ONPG (O-nitrophenyl-β-D-galactopyranoside), and fluorescence under long-wavelength UV light when the substrate MUG (5-methylumbelliferyl-β-D-glucuronide) is metabolised by *E. coli* containing β-glucuronidase. The analytical method involves adding commercial dried indicator nutrients containing the two defined substrates to a 100 ml volume of water and incubation at 35-37°C (APHA, AWWA, WEF 1998). The result is either presence/absence testing in the 100 ml volume or quantification in a proprietary tray (QuantiTray™) which separates the sample into a series of test wells and provides a most probable number per 100 ml of water.

Table 8.4 shows some regularly used chromogenic substances available for the detection of indicator bacteria.

**Table 8.4. Examples of chromogenic substrates for the detection of indicator bacteria**

(Adapted from Manafi, 1996)

| Bacteria | Chromogenic substance | Enzyme tested |
|---|---|---|
| Coliform bacteria | o-nitrophenyl-β-D-galactopyranoside (ONPG)<br><br>6-bromo-2-naphtyl-β-D-galactopyranoside<br><br>5-bromo-4-chloro-3-indolyl-β-D-galactopyranoside (XGAL) | β-D-galactosidase (E.C.3.2.1.23) |
| E. coli | 5-bromo-4-chloro-3-indolyl-β-D-glucuronide (XGLUC)<br><br>4-methylumbelliferyl-β-D-glucuronide (MUG)<br><br>p-nitrophenol-β-D-glucuronide (PNPG) | β-D-glucuronidase (GUD) (E.C.3.2.1.31) |
| Enterococci | 4-methylumbelliferyl-β-D-glucoside (MUD)<br><br>indoxyl-β-D-glucoside | β-D-glucosidase (β-GLU) (E.C.3.2.21) |

A major concern with any assay based on enzyme activity, is the interference that can be caused by the presence of other bacteria. In addition, the use of β-galactosidase in coliform detection has other disadvantages, as the enzyme can be found in numerous organisms (including *Enterobacteriaceae, Vibrionaceae, Pseudomonadaceae* and *Neisseriaceae*, several Gram-positives, yeasts, protozoa and fungi).

β-glucuronidase activity although produced by most *E. coli* strains is also produced by other *Enterobacteriaceae* including some *Shigella, Salmonella, Yersinia, Citrobacter, Edwardia* and *Hafnia* strains. The presence of this enzyme in *Flavobacterium* spp., *Bacteroides* spp., *Staphylococcus* spp. *Streptococcus* spp., anaerobic corynebacteria and *Clostridium* has also been reported. This could lead to the detection of a number of false positive organisms. On the other hand, some strains of *E. coli* (among them pathogenic strains) cannot be detected with this technique since they are (phenotypically) β-glucuronidase negative. Nonetheless, the above problems generally result in fewer errors than traditional cultivation-based methods.

*8.3.1.2   Cultivation in host cells*

*Cultivation of phages*

Numbers of phages are generally determined by direct quantitative plaque assays, the principles of which were designed by Adams (1959). Basically, soft agar is mixed with a suitable volume of the water under investigation and a culture of the host bacterium of choice at a temperature just above the solidification temperature of the agar. This mixture is poured on top of a bottom agar in a conventional 90 mm diameter Petri dish, yielding what is called a double agar layer (DAL) assay. The plates are incubated and plaques scored the following day. In tests on waters containing high numbers of bacteria (which may interfere with the host strain and the visibility of plaques) antibiotics such as nalidixic acid may be added to the agar medium and a resistant host strain is used.

A significant limitation of DAL methods is that only approximately 1 ml of test water can be used per 9 cm Petri dish. The quantitative detection of phages in numbers below the detection limit of direct plaque assays is, therefore, carried out by direct plaque assays using large Petri dishes, or the recovery of phages from large volumes of water followed by conventional plaque assays on the concentrates. Small numbers of phages in large volumes of water may also be detected by qualitative enrichment procedures.

*Cultivation of viruses*

The detection of viruses following the concentration step is performed in flat-bottom stationary flasks/wells or in rotating test tubes (roll-tubes) containing specific cell lines. Viruses are thus counted as plaques (clearings) in solid monolayers of cells, as tissue culture for 50% infective dose ($TCID_{50}$) or most probable number (MPN) in liquid suspensions (Payment, 2001).

*Monolayer Plaque Assay*: The cultivable enteroviruses produce a characteristic cytopathic effect and some can also produce visible plaques under a solid nutrient overlay. For the detection of plaque-forming enteroviruses, the plaque assay has been widely used. It has the advantage of providing results for rapidly growing viruses and can provide an isolated plaque (the equivalent of a bacterial colony), which can be picked up and contains a single virus type useful for virus identification and propagation. Disadvantages include under estimating the number of slow-growing viruses and not being able to detect those that are not plaque-producing.

254

A draft European Committee for Standardisation document (CEN/TC230/WG3/TG4) describes the monolayer plaque assay for enterovirus as follows: Confluent monolayer of Buffalo Green Monkey (BGM) cells in flasks or cell culture grade dishes are inoculated with the sample and incubated for one hour at 37°C (+/- 1.5°C). Excess sample is removed and an overlay medium containing agar and neutral red is added and allowed to set. After incubation pale areas of cell death (plaques) develop and are counted up to seven days. The cytopathic effect is localised because the agar will only allow spread of virus from cell to cell and neutral red is only taken up by living cells. It is assumed that a plaque is the progeny of a single virus infectious unit and they are referred to as plaque forming units (pfu). The number of pfu in the original sample can be calculated, utilising whole or part of the sample concentrate in multiple assays.

*Liquid overlay assays*: Slow-growing and non-plaque producing enteroviruses as well as viruses from the other groups (adenoviruses, reoviruses, hepatitis A, rotaviruses, etc.) replicate in cells but do not always produce any microscopic changes. To increase the probability of finding the viruses, one, two or even three passages incubated for seven to 14 days increase the probability of virus detection by allowing several cycles of replication. These techniques, under liquid nutrient medium can be performed in a macro-technique (tubes or flasks) or, more commonly, in a micro-technique (multiwell plates: 96, 24, 12, six or four wells). The number of inoculated tubes or wells determines the precision of the assay. When testing samples with a probable low number of viruses, a small number of flasks with a large surface area is preferable in order to maximise isolation and reduce the required labour time. The assay relies on various detection methods to enumerate the viruses in the original samples, including:

- Cytopathic effect (microscopy).

- Immunofluorescence (with specific or group antisera).

- Immunoperoxidase (with specific or group antisera).

- Molecular methods (PCR, hybridisation, etc.).

- Detection of virions in the supernatant by electron microscopy.

- ELISA methods (specific for one or more viruses).

Examples of frequently used cell lines are: MA 104, BGM-Fl, BGM-H, RD, Frhk 4, HFS, HEP, Vero, CaCo-2.

*Cultivation of protozoa in cell culture*

In contrast to most bacteriological and virological assays, parasitological (protozoological) samples do not incorporate an enrichment step based on *in vitro* cultivation of the captured organisms in general. Improved *in vitro* assays for *Cryptosporidium parvum* have been developed to demonstrate the infectivity of the parasite. The majority of the life cycle can be completed in tissue culture but the production of new oocyst numbers is low and usually less than that used for the inoculum.

The methods for the cultivation of *C. parvum* may serve as an example for other protozoa (such as *Toxoplasma gondii, Isospora belli, Cyclospora cayetanensis* and various genera of Microsporidia). A variety of cell lines (*e.g.* CaCo-2 cells, bovine fallopian tube epithelial cells, Mardin Darby Bovine Kidney cells, HCT-8 cells) are currently in use for the cultivation of *C. parvum* (Slifko *et al.*, 1997; Gasser and Donaghue, 1999). One typical cultivation method is outlined below. *C. parvum* oocysts are treated with 10% bleach (5.2% sodium hypochlorite, or the sporozoites freshly recovered by the process of the excystation) and plated onto HCT-8 cells grown to approximately 60 to 80% confluency in a 5% $CO_2$ atmosphere at 37°C. Oocyst formation can be detected three days after inoculation. Propagation in cell cultures may be used in combination with polymerase chain reaction (see Box 8.3) for the detection of infectious oocysts, however, it requires specific staff training and experience and special equipment.

*Cultivation of protozoa on artificial media*

Artificial culture media for both *Entamoeba histolytica* and *Giardia lamblia* have been developed and used for diagnosis in the medical field. Historically, these lumen-dwelling protozoa have been grown in culture media with and without one or more of the microorganisms with which they are associated in their normal habitat within the hosts (xenic culture). Cultivation techniques so far developed are not quantitative and have never been successfully applied to environmental samples.

### 8.3.1.3    Standardisation of methods

Established standard methods are available, *e.g.* those of the International Organization for Standardization (ISO), the European Committee for Standardisation (CEN) and the American Public Health Association (APHA). Methods for the detection and enumeration of indicator bacteria and some pathogenic or opportunistic bacteria are so widely needed that international standardisation is well underway. Table 8.5 outlines the state of the art of international standardisation of microbiological methods relevant for drinking water analysis.

**Table 8.5. International standardisation of methods for microbiological drinking water analyses**

ISO numbers refer to a published standard, () standard proposal not yet published or [] published standard under revision

| Target organisms | ISO standard | Culturing technique, medium/media and incubation | Observations |
|---|---|---|---|
| Coliform bacteria, Thermotolerant coliforms, *Escherichia coli* | [ISO 9308-1] | Membrane filtration on a selective medium; incubation (after resuscitation) at 36°C for 1 day (coliforms) or at 44°C for 1 day (thermotolerant coliforms); for confirmation of coliforms subculturing for lactose fermentation and gas production at 36°C for 2 days; for confirmation of thermotolerant coliforms subculture for lactose fermentation and gas production at 44°C for 1 day; for confirmation of *E. coli* subculturing for indole production at 44°C for 1 day is additionally needed; oxidase test | Poor selectivity; target colonies difficult to score |
| Coliform bacteria, Thermotolerant coliforms, *Escherichia coli* | [ISO 9308-2] | Liquid culturing in a selective medium; incubation at 36°C for 2 days; for confirmation (gas production) subculturing in BGBB at 36°C for 2 days for coliforms and in EC medium at 44°C for 1 day for thermotolerant coliforms, and additionally testing for indole production at 44°C for 1 day; oxidase test | A choice from several selective media is allowed in this outdated standard; revision is delayed due to lack of validation data on different media, coliforms as the target group taxonomically too heterogenic; time-consuming enumeration; material not expensive but labour costs significant |

258

**Table 8.5. International standardisation of methods for microbiological drinking water analyses (*continued*)**

| Target organisms | ISO standard | Culturing technique, medium/media and incubation | Observations |
|---|---|---|---|
| Faecal enterococci | ISO 7899-2 | Membrane filtration on m-enterococcus agar, incubation at 36°C for 2 days; *in situ* aesculin hydrolysis test on BEAA at 44°C for 2 hours | Ecology of the target group should be re-evaluated due to new taxonomy; time consuming enumeration; material not expensive |
| Faecal enterococci | [ISO 7899-1] | Cultivation in liquid medium, azide glucose broth, incubation at 36°C for 1 and 2 days; subculturing on BEAA at 44°C for 2 days; catalase test | Ecology of the target group should be re-evaluated due to new taxonomy; time consuming enumeration; material not expensive |
| Sulphite-reducing clostridia, spores | [ISO 6461-2] | Normal or modified membrane filtration on sulphite-iron or tryptose-sulphite agar anaerobically at 37°C for 1 and 2 days | Pasteurisation of the sample enhances spore germination as well as their selection by killing vegetative cells; the target group poorly defined; material costs increase if anaerobic jars are used |
| Sulphite-reducing clostridia, spores | [ISO 6461-1] | Liquid culturing in DRCM anaerobically at 37°C for 2 days | Pasteurisation of the sample enhances spore germination as well as their selection by killing vegetative cells; the target group poorly defined; material not expensive |
| *Pseudomonas aeruginosa* | [ISO 8360-2] | Membrane filtration on Drake's medium 19, incubation at 37°C for 2 days; for confirmation subculturing on milk agar at 42°C for 1 day (growth, casein hydrolysis, fluorescence and pyocyanine) | Atypical isolates should be further identified; material not expensive but labour costs significant; revision carried out within CEN |

**Table 8.5. International standardisation of methods for microbiological drinking water analyses** *(continued)*

| Target organisms | ISO standard | Culturing technique, medium/media and incubation | Observations |
|---|---|---|---|
| *Pseudomonas aeruginosa* | [ISO 8360-1] | Liquid culturing in Drake's medium 10 at 37°C for 2 days; for confirmation subculturing on milk agar at 42°C for 1 day (growth, casein hydrolysis, fluorescence and pyocyanine) | Atypical isolates should be further identified; material not expensive but labour costs significant |
| *Legionella species* | ISO 11731 | Spread plating on GVPC medium with antibiotics at 36°C for 10 days; subculturing on BYCE and BCYE-cys; serological testing of isolates growing on BYCE but not on BCYE-cys ; identification by fatty acids, isoprenoid quinones, indirect or direct immunofluorescent antibody assay, slide or latex bead agglutination, genus-specific monoclonal antibody or enzyme-linked immunosorbent assay | With and without sample pretreatment; background growth interferes; antibiotics and identification increase costs |
| *Legionella species* | (ISO/DIS 11731-2) | A screening method based on membrane filtration | |

**Table 8.5.  International standardisation of methods for microbiological drinking water analyses** *(continued)*

| Target organisms | ISO standard | Culturing technique, medium/media and incubation | Observations |
|---|---|---|---|
| *Salmonella species* | [ISO 6340] | Liquid pre-enrichment in buffered peptone water at 36°C for 1 day, enrichment in modified Rappaport-Vassiliadis broth at 42°C for 1 day, selection on brilliant green/phenol red lactose and xylose lysine deoxycholate agar at 36°C for 1 day and optionally on bismuth sulphite agar at 36°C for 2 days; isolation of typical colonies for confirmation using biochemical and serological tests | *S. typhi* needs another pre-enrichment medium; time and many media needed which increases costs |
| Culturable microorganisms | ISO 6222 | Pour plate technique, yeast extract agar, incubation at 36°C for 2 days and at 22°C for 3 days | All microorganisms are not expected to generate colonies; changes in colony forming units (cfu) relevant; cheap method |
| F-RNA phages | ISO 10705-1 | Double layer pour plate, *Salmonella typhimurium* strain WG49, phage type 3 Nalr (F' 42 *lac*::Tn5): NCTC 12484; *Escherichia coli* K12 Hfr: NCTC 12486 or ATCC 23631 as host bacteria, TYGB and TYGA at 37°C for different periods depending on step; with and without RNase | AQC on the host bacterium necessary, count of RNA phages on the basis of subtraction of DNA phages from total plaque forming units (pfu); RNA phages produce small plaques |
| Somatic coliphages | ISO 10705-2 | Double layer pour plate, ATCC 13706 strain as host bacterium, MSB and MSA at 37°C for different periods depending on step | Most sensitive of all the phage methods; multiplication of somatic coliphages possible, but appears not to be significant. |

261

**Table 8.5. International standardisation of methods for microbiological drinking water analyses** *(continued)*

| Target organisms | ISO standard | Culturing technique, medium/media and incubation | Observations |
|---|---|---|---|
| *Bacteroides fragilis* phages | (ISO 10705-4) | Double layer pour plate, defined strain as host bacterium, BPRM at 37°C for different periods depending on step, incubation anaerobically | Low numbers compared with somatic coliphages; host bacterium fastidious |
| Concentration of bacteriophages | (ISO/DIS1070 5-3) | Validation method described | |
| Enteroviruses | CEN/TC 230 | Cultivation on mono-layer of BGM cells at 37°C for 7 days; 5% of $CO_2$ in the atmosphere | Detects a range of enteroviruses, but BGM cells may be inhibited by components from environmental samples |
| Evaluation of membrane filters | ISO 7704 | Comparison of relative recoveries for a method | |
| Evaluation of colony count media | ISO 9998 | Comparison of relative recoveries for a method | |
| Validation of microbiological cultivation methods | (ISO TR 13843) | Characterisation of methods and confirmation of the detection of the target organism | |
| Equivalence testing of microbiological cultivation methods | (ISO/CD 17994) | Comparison of relative recoveries of target organisms between different methods | |

Key: AQC: Analytical Quality Control. ATCC: American Type Culture Collection (Manasses, VA, USA). BCYE: Buffered Charcoal Yeast Extract agar. BCYE-cys: BCYE without cysteine. BEAA: Bile Esculin Azide Agar. BGBB: Briliant Green Bile Broth. BGM: Buffalo Green Monkey. BPRM: Bacteroides Phage Recovery Medium. DRCM: Clostridium Differential Bouillon. EC: selective broth for coliforms and E. coli (containing bile salts). GVPC: BCYE with Legionella selective supplements. MSA: Modified Scholtens' Agar. MSB: Modified Scholtens' Broth. NCTC: National Collection of Type Cultures and Pathogenic Fungi (Public Health Laboratory Service, UK). TYGA: Tryptone Yeast Extract-Glucose Agar. TYGB: Tryptone Extract-Glucose Broth.

## 8.3.2    Detection and identification

Molecular methods targeting nucleic acids are the necessary tools for unveiling microbial diversity and can be used in detection and identification. Basic nucleic acid methods are hybridisation, restriction, amplification, cloning, and sequencing and these are summarised in Box 8.2.

---

**Box 8.2. Molecular methods targeting nucleic acids**

**Hybridisation** is a reaction involving two complementary nucleic acid strands, which bind to form a double-stranded molecule. Often, one of the nucleic acids is denatured total bacterial DNA and the other is a nucleic acid fragment used as a probe. The probe is either a cloned DNA fragment or, more conveniently, a synthetic oligonucleotide (usually 15 - 25 nucleotides long). The hybridisation reaction can easily be followed when the probe is labelled. Non-radioactive label is visualised by an immuno-enzymatic reaction or a cascade of reactions involving avidin and a biotinylated enzyme, when the label is biotin. Visualisation is colorimetric, fluorescent, or luminescent. Alternatively, oligonucleotide probes can be bound to a support (filter, microtiter plate, micro-chip) and unknown DNA labelled.

Some parameters (temperature, ionic strength) must be controlled in order for the hybridisation to work properly (Grimont, 1988). For given reacting sequences and ionic strength, there is an optimal temperature allowing maximum binding of the probe. A stringent temperature allows the best removal of incompletely reassociated nucleic acid while retaining enough perfectly bound probe to allow the detection of an unambiguous signal (*e.g.* colour). A low ionic strength allows the use of a lower temperature. Optimal (or stringent) temperature and ionic strength depend on the length of perfectly hybridised nucleic acid. This means that optimal temperature for a given probe will allow some partial hybridisation (cross-reaction) when probe and target sequences do not match perfectly (heterologous nucleic acid). For a given probe, the specificity index is defined as the dilution needed to lower the homologous reaction to a level similar to that of a heterologous reaction (Grimont *et al.*, 1985). In other terms, excessive numbers of non-target bacteria may give false positive results. This is probably why hybridisation works well for confirmation of culture identification (when nucleic acid amounts are controlled) but is often inconclusive when used on field samples with unknown amounts of nucleic acid or mixtures of unknown numbers of bacterial species. It should also be noted that probes targeting DNA do not differentiate between live and dead bacteria.

---

## Box 8.2. Molecular methods targeting nucleic acids *(continued)*

**Restriction endonucleases** are enzymes that recognise short specific palindromic sequences and cleave double stranded DNA at these sites. Digesting a DNA molecule with a given restriction endonuclease yields a finite number of DNA fragments. Electrophoresis is used to separate restriction fragments by size. Restriction of a bacterial genome often produces too many fragments to be analysed. This problem has been solved in two ways: one way is to use restriction enzymes that recognise rare cleavage sites thus generating a few very large fragments. The latter can be separated with the help of a special technique referred to as pulse field gel electrophoresis. The other way is to visualise a subset of fragments after hybridising with a specific probe (Southern method). When the probe targets 16 and 23S rRNA genes, the method is applicable to all bacteria and is often referred to as ribotyping (Grimont and Grimont, 1986). Alternatively, a DNA fragment can be amplified (see below) and digested by a restriction endonuclease to give a simple pattern (Kilger and Grimont, 1993). Restriction methods are best applied to purified DNA extracted from pure culture (bacterial identification and typing) and are not currently used on field samples with complex bacterial flora.

**Amplification** is a method in which a chosen nucleic acid sequence (DNA or RNA) is copied many times. Currently, the polymerase chain reaction (PCR) is the most widely used principle (see Box 8.3). A major problem with this technique relates to the high sensitivity of PCR, which allows the amplification of contaminating polynucleotides when careful procedures are not implemented. Furthermore, dead bacteria (*e.g.* autoclaved or disinfected) can still be detected by PCR. On the other hand, sample specific substances can interfere with the PCR reaction and may seriously affect the detection limit (Wilson, 1997).

The above methods can be combined. Restriction and hybridisation are used by the Southern method. Restriction of amplified products are used for identification (when rRNA genes are amplified) or typing (*e.g.* flagellin genes). Selective amplification of restriction fragments is used in the method called Amplified Fragment Length Polymorphism (AFLP).

An emerging technology consists of using arrays of probes bound to a support (membrane or microchip). Amplified target DNA is hybridised with the bound probes and individual reactions are scored either using some electronic device or by image analysis. Although the molecular techniques used are not new (Rijpens *et al.*, 1995), and unable to distinguish between live and dead bacteria when DNA is targeted, probe multiplicity (several thousand) and miniaturisation are interesting in many fields, especially the identification of alleles of many genes in a given bacterial strain.

**Cloning** is a method in which a restriction fragment is inserted in an autoreplicating vector (plasmid, phage, and cosmid) and thus biologically amplified.

**Sequencing** often uses a cloned or amplified gene and oligonucleotides (which hybridise to part of the gene), DNA polymerase (which copies the gene) and nucleotide analogues (which randomly stop elongation when adding a given nucleotide type). The result is a family of fragments all ending with a given nucleotide type. These are separated by electrophoresis. Sequences can be read automatically and compared with those contained in databases. Several databases are available on the Internet.

*8.3.2.1   Polymerase chain reaction (PCR) – based detection*

## *Polymerase chain reaction (PCR)*

A basic laboratory infrastructure is essential to perform PCR (see Box 8.3). Various kits are commercially available from different suppliers which provide all protocols and reagents needed to carry out PCR based assays. In addition, a thermocycler for the PCR reaction and appropriate equipment for separation (*e.g.* power supplies, electrophoresis units) and detection/visualisation of nucleic acids are required.

---

**Box 8.3. Amplification of nucleic acids**

Among a variety of nucleic acid amplification technologies the **Polymerase Chain Reaction** (**PCR**) is the most prominent example. This method, carried out *in vitro* in a thermocycler, takes advantage of the thermostability and fidelity of DNA polymerases from certain thermophilic bacteria. Using DNA probes as templates and two oligonucleotide primers that bind to complementary sequences flanking the target, this method allows the exponential multiplication of nucleic acid fragments, in the presence of deoxynucleotides, within several hours. PCR amplification of DNA occurs in three steps: denaturation, annealing and extension of the primers. From the analysis of such amplicons, which can be separated in an electrophoretic step, it is possible to draw conclusions on the microorganisms that were present in the original sample (*e.g.* water). Thus, depending on the target sequence and the choice of primers, this technology allows the indirect detection of large groups of organisms or alternatively the identification of specific (sub)species. This can be achieved by using primers that bind to more conserved targets (*e.g.* regions encoding 16S rRNA) or to regulatory DNA sequences or genes that might be associated with specific functions such as virulence determinants (*e.g.* streptococcal pyrogenic exotoxins), respectively.

In order to prove that the amplified sequence is indeed the desired target a second PCR step can be included. **Semi-nested PCR** is performed with one external primer (used in the original amplification) and one internal primer that is designed from sequences contained in the expected first amplicon. In **nested PCR** two internal primers are used. This double PCR technique can increase the sensitivity of detection by one to two orders of magnitude.

**Duplex PCR** and **multiplex PCR** involve the use of two or multiple sets of primers resulting in two or multiple amplification products. When all products of amplification are diagnostic for a particular species it is possible to distinguish between closely related (sub)species. Alternatively, more than one organism can be detected when primers are used that are specific to different genomes.

Introducing molecular "tags", such as digoxigenin (DIG) or biotin-labelled dUTP into the PCR product can provide an invaluable tool for diagnostics. Such labelled PCR products may either be used as hybridisation probes or be detected by use of capture probes. For instance, with PCR-generated DIG-labelled hybridisation probes, it is possible to detect and quantify minute amounts of a pathogen.

---

**Box 8.3. Amplification of nucleic acids** *(continued)*

As DNA can survive for long periods after cell death the PCR reaction does not distinguish between viable and non-viable organisms. In contrast to DNA, messenger RNA (mRNA) which is transcribed from DNA is very labile with a typical half-life of only a few minutes. A technique that offers potential for assessing viability and the presence of mRNA is **reverse transcriptase (RT) PCR.** There are two steps:

- Reverse transcription which produces DNA fragments from RNA templates, and

- PCR, which produces multiple copies of the target DNA.

In the first step a reverse transcriptase enzyme is used to extend an oligonucleotide primer hybridised to a single-stranded RNA containing the message of interest, producing a complementary DNA strand (cDNA). DNase I is used to remove any contaminating DNA that may result is false positive results. The entire process usually takes around three hours. Total RNA, messenger RNA (mRNA), transfer RNA (tRNA) or ribosomal RNA (rRNA) from a variety of sources (bacteria, viruses, parasites, yeast, plants, etc.) can be used as templates for reverse transcription. As in PCR, the amplified DNA segments corresponding to the target sequences can be detected using standard detection methods such as agarose gel electrophoresis or membrane hybridisation with specific DNA probes or by using ELISA.

**Random Amplified Polymorphic DNA (RAPD)** or **Arbitrarily Primed PCR (AP-PCR)** is a specialised form of PCR. It differs from normal PCR in that a single short primer (generally ten base pairs long) of a random sequence is utilised for the amplification of genomic DNA. This single short primer can anneal randomly at specific sites within a genome. Priming sites are randomly distributed throughout a genome and polymorphisms in such sites result in differing amplification products, detected by the presence or absence of fragments.

In principle, nucleic acid might be detected by PCR from all waterborne viruses and (micro)organisms, as long as their envelopes (capsids, membranes, cell walls) can be disrupted to make the nucleic acids accessible to the enzymatic reaction. For the release of nucleic acids from viruses and microorganisms different methods such as freeze-thaw cycles, boiling, addition of detergents, digestion with enzymes are applied.

The complete procedure including sample preparation is fast compared to the 'classical' methods, with results available in three to four hours. The amplified DNA fragments can be easily detected by gel electrophoretic separation and subsequent staining techniques, and could be analysed further. PCR is very flexible, and allows highly specific detection of particular (sub)species (*e.g. Escherichia coli*: EHEC, ETEC, STEC, UPEC), certain groups of microorganisms (*e.g. Enterobacteriaceae*) or can be used to study aspects of biodiversity in water samples. Since the nucleic acid region that is flanked by the primers does not have to be known completely, uncultured pathogenic microorganisms might be discovered in broad-range PCR approaches.

PCR techniques also have limitations. Although PCR is very sensitive samples have, in most cases, to be concentrated (*e.g.* by flocculation, filtration, centrifugation, precipitation, immuno magnetic separation, adsorption to particles followed by sedimentation and elution). The method may generate false positive results, especially when carried out without a pre-cultivation step of the original water sample, since it does not discriminate between viable and non-viable organisms. In addition, DNA molecules can survive in the environment for long periods of time, this may also result in a number of false positive reactions. The basic procedure does not allow quantitation of the number of the amplifiable DNA (RNA) fragments in the original sample. The PCR reaction is sensitive to inhibition by compounds that are present in environmental water samples (*e.g.* divalent cations, fulvic and humic acids) and might vary depending on thermocyclers and reagents used from different suppliers. Despite these limitations, in the coming years PCR and PCR-based methods are likely to be further automated, allow quantitation and be used in routine laboratories.

### *Reverse transcriptase (RT – PCR)*

The indication of the viability of microorganisms in a given sample would be of enormous significance for food, industrial, environmental and medical applications. RT-PCR is useful to detect the presence of specific messenger RNA (mRNA) or ribosomal RNA (rRNA) sequences (see Box 8.3). Messenger RNA is turned over rapidly in living bacterial cells. Most mRNA species have a half-life of only a few minutes (Belasco, 1993). Detection of mRNA by RT-PCR might therefore be a good indicator of living cells or those only recently dead at the time of sampling (Sheridan *et al.*, 1998). This method has been used to determine the viability of *Legionella pneumophila*, *Vibrio cholerae* as well as that of *Giardia* cysts and *Cryptosporidium* oocysts through the detection of heat shock protein hsp70 mRNA (Bej *et al.*, 1991; Bej *et al.*, 1996; Stinear *et al.*, 1996; Abbaszadegan *et al.*, 1997) in environmental water samples. The method, combined with calcium carbonate flocculation to concentrate samples, induction of hsp70, and purification by immunomagnetic separation has been shown to be able to detect a single oocyst. However there are still disadvantages using this technology. The method is qualitative not quantitative and as the oocysts are broken subsequent counting is not possible.

Despite their large advantages, RNA based approaches face technical difficulties, particularly the extraction of detectable levels of intact RNA (a molecule that is significantly less stable than DNA). In order to minimise this problem a number of commercial kits for extraction and purification of RNA have been developed. The enzyme, reverse transcriptase, like the polymerases for PCR is highly susceptible to a number of inhibitory contaminants commonly found in water (*e.g.* humic compounds). Therefore, considerable efforts have to be made in order to remove these compounds prior to testing. Immunomagnetic capture, as well as nucleic acid capture have proven to be successful for this purpose. Oligonucleotide probe-linked magnetic beads combined with RT-PCR have been used for the detection of viable *Giardia* and *Cryptosporidium* in water samples containing PCR-inhibiting substances. Although, as in the case of PCR, loss of microorganisms during the concentration and recovery can greatly reduce the detection sensitivity of the method.

Direct PCR does not distinguish between infectious and non-infectious viral sequences. The integrated technique involves inoculation of the concentrated sample onto cell monolayers, which are then incubated for a minimum of 24 hours. This allows virus RNA to be amplified in tissue culture making RT-PCR on cell culture lysate more sensitive. The technique is known as integrated cell culture RT-PCR (ICC RT-PCR).

Quantification using RT-PCR is still difficult, laborious and inaccurate, and requires skilled operators and large amounts of materials. Because neither PCR nor RT-PCR provides reliable means for quantification, commonly RT-PCR detection of pathogens in water has only been used as a qualitative presence/absence test. In recent years advances in technology and products have been made towards quantification of PCR and RT-PCR. These developments, (*e.g.* the TaqMan™ and LightCycler™ systems) are very promising for 'in-tube' detection and quantification.

### 8.3.2.2    *Fluorescence in situ hybridisation (FISH)*

With the help of *in situ* hybridisation techniques organisms can be detected in their natural habitat without the need for pre-culture techniques. The method involves fixation of the cells in their natural state followed by permeabilisation of the cell wall. This enables all the reagents, including the species-specific oligonucleotide probes, to move into the cell and hybridise to their target. The probes are labelled with fluorescent dyes to enable the hybridised target within the cell to be viewed by epifluorescence microscopy or laser scanning electron microscopy. Fluorescent oligonucleotide probes can react with, for example, all bacteria of a given phylogenetic branch, a genus, or a single species (Amann *et al.*, 1990, 1995). Different fluorescent labels can be used enabling multicolour reactions. Possible targets for hybridisation are genes, mRNA and rRNA. Genes cannot be detected by *in situ* hybridisation unless some *in situ* PCR step is used, as some 10 000 labelled molecules are typically required for 'visualisation'.

## Detection of bacteria by FISH

A typical *in situ* hybridisation protocol includes filtration of a water sample through a membrane, fixation of bacterial cells on the membrane, permeation of cells (to allow the probe to access its target), hybridisation with a fluorescent probe, washing to eliminate unbound probe, and microscopic examination.

rRNA molecules are universally present in bacteria, have diversely conserved portions of their sequences, and occur in about 30 000 copies per 'active' cell, as such they are perfect targets for *in situ* hybridisation. Databases contain many rRNA sequences. However, since sequences are not available for all described species, probes must be tested against a collection of reference microorganisms.

The fluorescent signal given by a cell depends on the number of targets and the accessibility of the target. For a given probe targeting rRNA, the signal is in relation to the number of ribosomes (*i.e.* the physiological state of the bacterium). When a probe is designed, attention should be given to target accessibility as the fluorescent signal varies widely depending on the sequence position of the target on the rRNA (Fuchs *et al.*, 1998). The fluorescent signal is stronger when the probe is longer with multiple labels (Trebesius *et al.*, 1994) or when a peptide nucleic acid (PNA) is used as a probe (Prescott and Fricker, 1999). However, for a given probe sequence, PNA probes are less specific than regular oligonucleotide probes and mismatches must be introduced to raise specificity. Signal amplification systems, such as Tyramide Signal Amplification (TSA) accumulate fluorescent compounds in cells where a probe reacted. This gives a very strong signal and may allow detection and enumeration of fluorescent cells by scanning cytometry. PNA probes and TSA have been combined, and the use of PNA probes targeted against the 16S rRNA molecule for the specific detection of *E. coli* have been shown to offer a fast efficient alternative to conventional approaches (Prescott and Fricker, 1999). In this method, bacteria are captured by filtration of the water samples through metallic membranes. The cells are then fixed by placing the membranes on a filter pad pre-soaked with paraformaldehyde solution. They are then treated with lysozyme, washed and overlaid with hybridisation solution containing a biotinylated PNA oligonucleotide probe specific for the detection of *E. coli*. The biotin PNA-RNA complex is detected by incubation in streptavidin horseradish peroxidase (HRP) followed by the addition of fluorescein tyramide. The HRP catalyses the deposition of fluorescein and the cells are detected by epifluorescence microscopy. The test requires no specialised equipment and is easy to perform in two to three hours. The procedure could be performed directly on the water sample without the need for culture techniques. Unfortunately, dead bacteria can also be detected after signal amplification.

Problems and shortcomings have been identified when FISH is applied to the detection of bacteria in water:

- **Detection is strictly taxonomic.** Molecular detection and identification of bacteria work in the framework of molecular taxonomy. Taxonomic groups that are not confirmed by molecular methods may not be properly identified. As an example, coliforms (whether faecal or not) do not constitute a taxon in molecular terms, therefore no nucleic acid probe can detect them. It is possible to use a probe or PCR system targeting the beta-galactosidase gene. Such probes, however, will not react with all coliforms but rather with coliform species that are phylogenetically close to *E. coli*, irrespectively of their habitat. Moreover, all *Shigella* species and serotypes, except *S. boydii* 13, belong to the *E. coli* genomic species and *Shigella* spp. are seen as entero-invasive clones of *E. coli*. Therefore, no taxonomic probe can distinguish *Shigella* spp. from *E. coli*. Probes (or PCR systems) can target the invasivity genes. These probes will detect invasive strains of *Shigella* and *E. coli* but will not detect *Shigella* strains that have lost the invasivity plasmid.

- **Bacteria in water are often starved or stressed.** Starved or stressed bacteria are less reactive and often occur as tiny cells. As such, they are difficult to distinguish among some inanimate material which may bind probes non-specifically. Furthermore, naturally fluorescent bacteria or objects may occur. A major problem with molecular methods is to distinguish live from dead bacteria. A bacterial state has been described in which bacteria are viable but nonculturable (VBNC) (Roszak and Colwell, 1987; Colwell and Grimes, 2000). However, clear definitions of bacterial life and death are needed (Villarino *et al.*, 2000). When starved or stressed bacteria are incubated in the presence of nutrients, yeast extract, and nalidixic acid (or ciprofloxacin), the cellular machinery is restarted, ribosomes accumulate while the cell elongates (nalidixic acid or ciprofloxacin prevent cell division, not elongation), thus demonstrating viability. Such cells are easily differentiated from inanimate material and from dead cells. The method called Direct Viable Count, or DVC (Kogure *et al.*, 1979) was adapted to FISH (Kalmbach *et al.*, 1997; Nishimura *et al.*, 1993; Regnault *et al.*, 2000) and proved to be the most accurate viability marker (Villarino *et al.*, 2000). The DVC was also used to first detect VBNC bacteria in water (Xu *et al.*,1982).

- **The major drawback of microscopic methods is sensitivity.** In order to reach sufficient microscopic sensitivity to detect one cell per 100 ml, bacteria should be concentrated from volumes of 100 litres. Alternatively, machines scanning the whole filter surface for fluorescent objects could be used together with automatic positioning of the microscope above detected fluorescent objects.

## Detection of protozoa by FISH

Probes targeting the 18S rRNA and used in the hybridisation assay can be synthesised to the genus or species level. The use of FISH as an alternative technique would enable the specific detection of *Cryptosporidium parvum* as traditional methods such as antibody staining are unable to distinguish between different species within the genus (Vesey *et al.*, 1998). However fluorescence labelling does not produce fluorescence bright enough to be used for primary detection since autofluorescence particles like algae fluoresce more brightly. A combination of fluorescence and secondary antibodies should improve detection systems.

## 8.4    Emerging procedures

### 8.4.1    *Laser scanning analysis*

With the development of test procedures on the basis of chromogenic (Table 8.4) and fluorogenic substrates for the detection and enumeration of coliform bacteria and *E. coli,* the analysis can be performed in 24-48 hours. Several approaches have been investigated to enhance the sensitivity of enzymatic reactions using instrumentation instead of the traditional visual approach. Spectrophotometry has been shown to reduce the 24hr Colilert® test by 6 hours (Rice *et al.*, 1993). Using fluorometry one faecal coliform can be detected within 7 hours. In addition the 'ChemScan®' instrument from 'Chemunex®' has been used with membrane filtration tests for the detection of fluorescent microcolonies. Samples are filtered onto Cycloblack-coated polyester filters that are then incubated on a prefilter saturated with Colicult® medium. After the membrane has been transferred to a second prefilter saturated with a fluorogenic substrate the membrane is analysed by the ChemScan® instrument. The ChemScan® is a laser-scanning device with a motorised stage attached to an epifluorescence microscope. Each fluorescent event detected can be validated by microscopy. Initial experiments indicate that this method can be performed within 3.5 hours and yield results equivalent to those of standard methods. This technology allows the detection of any bacteria and protozoa that can be labelled with a fluorescent substrate linked to an antibody or nucleic acid probe. An alternative semi-quantitative system for the detection of coliforms is also available from Colifast®. This instrument called the CA-100 system also utilises the ability of coliforms to cleave galactoside conjugates to yield

271

fluorescent products. The level of fluorescence is measured at given time intervals and is directly proportional to the number of coliform bacteria present.

### 8.4.2 DNA – chip array

The future holds endless possibilities for the detection of both indicators and pathogens alike. On the horizon are methods based on microarrays and biosensors. Biosensors in the medical area have largely been based on antibody technology, with an antigen triggering a transducer or linking to an enzyme amplification system. Biosensors based on gene recognition, however, are looking very promising in the microarray format for detecting microorganisms. There are two variants of the DNA microarray technology, in terms of the property of arrayed DNA sequence with known identity:

• Probe cDNA (500~5 000 bases long) is immobilized to a solid surface such as glass using robot spotting and exposed to a set of targets either separately or in a mixture. This method was originally called DNA microarray and was developed at Stanford University (Ekins and Chu, 1999).

• An array of oligonucleotide (20~25-mer oligos) or peptide nucleic acid probes is synthesised either *in situ* (on-chip) or by conventional synthesis followed by on-chip immobilisation. The array is exposed to labelled sample DNA, hybridised, and the identity/abundance of complementary sequences are determined. This method, originally called GeneChip® arrays or DNA chips, was first developed at Affymetrix Inc. (Lemieux *et al.*, 1998; Lipshutz *et al.*, 1999).

Microarrays using DNA/RNA probe-based rRNA targets may be coupled to adjacent charged couple device detectors (Guschin *et al.*, 1997). Eggers *et al.* (1997) have demonstrated the detection of *E. coli* and *Vibrio proteolyticus* using a microarray containing hundreds of probes within a single well ($1cm^2$) of a conventional microtiter plate (96 well). The complete assay with quantification took less than one minute.

The microarray under development by bioMerieux (using Affymetrix Inc. GeneChip technology) for an international water company (Lyonnaise des Eaux, Paris, France) is expected to reduce test time for faecal indicators from the current average of 48 hours to just four hours. In addition, the cost for the standard water microbiology test is expected to be ten times less than present methods. The high resolution DNA chip technology is expected to target a range of key microorganisms in water. The prototype GeneChip® measures about 1 cm$^2$, on which hybridisation occurs with up to 400 000 oligonucleotide probes.

### 8.4.3 Biosensors

The biosensor relies on optics, immunoassays and other chemical tests, which may be directed to detect microorganisms. To date, most work has focused on bacterial pathogens (Wang et al., 1997). In general, there is an immunoaffinity step to capture and concentrate bacteria on beads, membranes or fibre optics probe tips, followed by detection by laser excitation of bound fluorescent antibodies, acoustogravimetric wave transduction, or surface plasmon resonance.

Several types of biosensors are currently under development, especially to detect foodborne pathogens in, for example, meat and poultry. As an example, one type is described (Georgia Tech Research Institute, 1999):

> The biosensor operates with three primary components - integrated optics, immunoassay techniques and surface chemistry tests. It indirectly detects pathogens by combining immunoassays with a chemical sensing scheme. In the immunoassay, a series of antibodies selectively recognise target bacteria. The 'capture' antibody is bound to the biosensor and captures the target bacteria as it passes nearby. A set of 'reporter' antibodies (which bind with the same target pathogen) contain the enzyme urease, which breaks down urea that is then added resulting in the production of ammonia. The chemical sensor detects the ammonia, affecting the optical properties of the sensor and signalling changes in transmitted laser light. These changes reveal both the presence and concentration of specific pathogens in a sample at extremely minute levels.

The method is currently unable to distinguish viable from non-viable microbes, and it will be necessary to increase the sensitivity in order to apply this technique to water testing. Nonetheless this methodology has a great potential for future application, especially as it is extremely fast.

### 8.4.4 Solid state biochips

The idea of rapid detection (minutes) of a number of toxins and actual microbial cells on a solid state biochip is a visionary approach currently being developed. This approach does not require isolation and characterisation of nucleic acids from the microorganisms and does not rely on capturing of antibodies. Further characteristics are: no lengthy incubation times, no labelling and no washing are needed. The technique is not yet available, so limitations cannot be determined.

## 8.5     Performance and validation of methods

### 8.5.1     Limitations and characteristics of microbiological methods

The low numbers of target organisms in microbiological and especially drinking water analyses increase measurement uncertainty. Even if assuming homogenous distribution of target in the sample, the numbers detected are defined by the Poisson distribution. Therefore, the uncertainty of measurement is related to each individual measurement result and method specific values are not satisfactory alone. To be of use in validation a clear specification should be available. This should include an exact description of the working conditions and media used, upper counting limit, recovery, working limits within which the method can be used, selectivity with the respect to the target organism, specificity, robustness and limitations of the method.

The aim of the selective recovery of target organisms from samples is challenging. In growth dependent methods the viability of a target organism is defined by growth of this organism under specified conditions (*i.e.* by the method itself, a non-selective method or a reference method). It is nearly impossible to determine the true number of viable target organisms that are present in a sample (even when the sample is spiked). Therefore, absolute recovery cannot be defined and for a new method only a relative recovery can be given by relating it to that obtained with other (reference) methods. Similar problems occur in molecularly based methods. Microbiological methods are not robust in the sense that chemical methods are. The target and many contaminants in the sample are living entities and therefore unexpected effects and phenomena can occur. Robustness is affected by many different factors including the physical, chemical and microbiological properties of sample itself. In the analyses based on nucleic acids, it has been repeatedly observed that sample specific inhibitors interfere and decrease sensitivity (Wilson, 1997). All the methods are affected by sample storage before analysis (*e.g.* cold-shocks), incubation conditions and the competence of the personnel executing the analysis (*e.g.* time needed to perform certain steps).

## 8.5.2    Statistical issues

In samples (even in well-mixed laboratory samples) particles, including target organisms, are unevenly distributed and this results in a random basic variation of the results obtained which cannot be avoided (Tillett and Lightfoot, 1995). Due to experimental imperfections the variations observed in practice in parallel determinations are even larger than predicted by the Poisson distribution (the mathematical law which describes the ideal distribution they should follow). This effect is called over-dispersion. In contrast to chemical analysis where even at low concentrations the number or target molecules in a sample volume is high this over-dispersion cannot be avoided in microbiological testing because the number of target microorganisms is usually low. It is therefore important to analyse a sufficient number of samples to obtain a convincing result. Although statistical theory provides clear information (*e.g.* Cochran, 1977; ISO/TR 13843, 2000) as to how many samples would be required for a certain testing scenario, it is often not possible to meet these requirements (*e.g.* because of cost reasons) and statistical considerations usually become guidelines only. However, generally it should be remembered that too few samples may be a waste of effort, time and money.

## 8.5.3    Validation of methods

Method validation provides evidence that a specific method is capable of serving the purpose for which it is intended (*i.e.* that it does detect or quantify a particular microbe [or group of organisms, or a viral particle] with adequate precision and accuracy). A new, or inadequately characterised, method is initially investigated in a primary validation process to establish its operational limits. Primary validation should result in an unambiguous and detailed quantitative description of the results the method can deliver. Primary validation of a new method is typically performed by the laboratory that has developed it. When the method is implemented in another laboratory secondary validation takes place (also referred to as verification). Here, it is established whether the specifications described in primary validation can be met. Usually only selected and simplified forms of the procedures used in the primary validation process are used, but over an extended period of time and/or more samples. It should be pointed out that validation should simulate the later routine as closely as possible and natural samples should be used as the main test material wherever possible.

For both primary and secondary validation it is of course essential that strict analytical quality control is used, because application of valid methods does not necessarily ensure valid results. The methods of analytical quality

275

control include, replications at different levels, inclusion of reference materials (qualitative and quantitative), intercalibrations and spiked samples (Lightfoot and Maier, 1988; McClure, 1990).

In case equivalent methods already exist, the justification for introducing a new method always requires careful comparison with one or more established methods in parallel on the same samples. Since every method usually consists of several steps, method performance includes many different aspects. For example, one method might be superior in specificity but inferior in recovery. One method might give highest recovery of target organisms but require confirmation of positive results in routine test. Hence, for routine use a method giving lower recovery but not requiring confirmation of positives is probably preferable. This indicates that it is frequently difficult to numerically specify the superiority of one method over the other.

Collaborative tests in which several laboratories participate are considered essential for the validation of microbiological methods as well as the performance of individual laboratories. These tools were developed for chemical analytical methods but many of the principles are now also applied to microbiological testing. These collaborative tests are mainly of two types (Horwitz, 1988; McClure, 1990):

- **Intercalibration exercises** which allow laboratories to compare their analytical results with those of other participating laboratories.

- **Method performance tests** that yield precision estimates (repeatability, reproducibility) when several laboratories analyse identical samples with strictly standardised methods. In such tests 'artificial' samples (*i.e.* certified reference materials and spiked samples) are included in the samples to be analysed by the participating laboratories.

Experience from chemical collaborative testing indicates that it is important that the participating laboratories have in-depth knowledge and experience with the methods to be tested and collaborative method performance tests are not used as laboratory proficiency tests and training exercises. It is important to note that a number of established microbiological methods (*e.g.* Endo agar for total coliforms or mFC for thermotolerant coliforms), although used for decades by hundreds of laboratories, have not been assessed in collaborative tests.

Whether or not a validated method is successful in practice may depend on political and/or commercial issues. For example, within ISO, acceptance and publication of a method as a standard method requires approval of at least 75%

of the member bodies casting a vote. On the other hand, stimulation of the use of certain methods by the development and promotion of easy to handle kits by commercial companies is also possible.

## 8.6    Summary

Table 8.6 summarises the predominant characteristics, advantages and limitations of the main detection methods that are described in this chapter.

**Table 8.6. Methods for the detection of microbial contamination in drinking water**

| Method | Characteristics/advantages | Limitations/disadvantages | Application: status quo and future perspective |
|---|---|---|---|
| Cultivation of bacteria | • Cultivation media mostly inexpensive<br>• Easy to perform<br>• Qualitative and quantitative results obtainable<br>• Differentiation and preliminary identification possible on selective solid media<br>• Detection of bacteria occurring in low numbers possible (in combination with concentration techniques, e.g. filtration) | • Time consuming<br>• Not all bacteria of interest can be cultivated<br>• Large sample volumes cause problems for some of the methods<br>• Does not detect 'viable but non-culturable' organisms<br>• Selectivity for the detection of certain indicators often not sufficient (false positive species)<br>• No information on infectivity of a pathogen<br>• Biosafety issues | • Standardised (ISO, CEN, APHA) for a number of species(groups)<br>• Improved media might be developed in order to obtain faster growth and to increase sensitivity and selectivity of the assays |
| Cultivation of bacterial viruses (bacteriophages) | • Assays inexpensive and easy to perform<br>• Quantitation possible<br>• Similar to bacterial methods<br>• Minimal biosafety issues (host cells) | • No direct correlation in numbers of phages and viruses excreted by humans<br>• Phages can be useful as faecal indicators, as well as models or surrogates for enteric viruses in water environments, but care is needed in interpreting the results. | • Standardised methods available (ISO) for major groups |

**Table 8.6. Methods for the detection of microbial contamination in drinking water** *(continued)*

| Method | Characteristics/advantages | Limitations/disadvantages | Application: status quo and future perspective |
|---|---|---|---|
| Cultivation of animal/human viruses | • Several enteric viruses can be propagated in cell culture (a variety of cell lines have been tested and used)<br>• Quantitation possible<br>• Growth indicates infectivity | • Requires some level of training and specialised laboratories<br>• Various cell lines may need to be used for the detection of a larger number of virus types<br>• Biosafety issues | • Standardised (ISO, CEN, APHA) for a number of species(groups)<br>• New cell lines are being developed and new media formulation may increase sensitivity |
| Cultivation of protozoa | • Excystation in vitro can be taken (to a certain extent) as indication for viability | • Does not provide information on infectivity for man<br>• Time consuming<br>• Propagation of most organisms in vitro using cell cultures is extremely poor<br>• Not all protozoa of interest can be cultivated<br>• Biosafety issues | • At present, the only available infectivity assay depends on animal hosts, which is costly and very time-consuming |
| Immunological detection of antigenic structures associated with microorganisms | • Qualitative and quantitative results regarding the number of microorganisms possible (to a certain extent)<br>• Relatively specific for target organism | • Often needs pre-cultivation step which is time consuming<br>• Lack of sensitivity<br>• Selectivity can be a problem due to cross-reacting antibodies<br>• Without pre-cultivation, currently no discrimination between viable and non-viable microorganisms<br>• No information on infectivity of a pathogen. | • Assays allow standardisation and automation |

**Table 8.6. Methods for the detection of microbial contamination in drinking water** *(continued)*

| Method | Characteristics/advantages | Limitations/disadvantages | Application: status quo and future perspective |
|---|---|---|---|
| Immunomagnetic separation (IMS) | • Faster and more specific than other concentration methods<br><br>• Sound basis for other detection methods (PCR, RT-PCR, FACS, FISH) as well as cultivation methods | • Sensitivity, robustness, consistency can be affected by environmental conditions<br><br>• Selectivity can be a problem due to cross reacting antibodies<br><br>• No information on infectivity of a pathogen | |
| Polymerase chain reaction (PCR) | • In principle highly sensitive (but see limitations)<br><br>• Selective<br><br>• Specific<br><br>• Can detect 'non-culturable' microbes<br><br>• Faster than cultivation methods (3-4 hours)<br><br>• Sound basis for further analyses of nucleic acids (sequencing, RFLP, RAPD)<br><br>• | • Limited reliability (at present the detection of an individual microbe cannot be guaranteed due to inconsistencies in performance of the technique)<br><br>• Sufficient quantity of nucleic acids from the targeted microbe has to be recovered<br><br>• Negatively affected by certain environmental conditions<br><br>• Basic procedure does not allow quantitation of the number of amplifiable DNA/RNA fragments<br><br>• At present no discrimination between viable and non-viable microorganisms<br><br>• No information on infectivity of a pathogen | • Currently no standardisation<br><br>• Potential for automation<br><br>• Potential for quantitation |

**Table 8.6. Methods for the detection of microbial contamination in drinking water** *(continued)*

| Method | Characteristics/advantages | Limitations/disadvantages | Application: status quo and future perspective |
|---|---|---|---|
| RT-PCR | • As PCR<br>• Good indication for living organisms with mRNA as target<br>• Can provide information on pathogenic potential of an organism when mRNA of a virulence gene is assayed | • As PCR (except discrimination between viable and non-viable microorganisms with mRNA as target)<br>• Extraction of detectable levels of intact RNA molecules is problematic due to their instability | • Currently no standardisation<br>• Potential for automation<br>• Potential for quantitation |
| Flow cytometry, fluorescence-activated cell sorting (FACS) | • Faster than cultivation methods<br>• Detection of non-culturable organisms | • No information on infectivity of a pathogen<br>• Expensive technology<br>• Limited reliability for the detection of microbes that are present in extremely low concentrations | |
| Fluorescence in-situ hybridisation (FISH) | • Faster than cultivation methods<br>• No pre-cultivation needed<br>• Detection of non-culturable organisms<br>• Can detect individual cells when ribosomal RNA is target<br>• Different (multicolour) fluorescent labels allow detection of different microbes<br>• Can be used in combination with machines that do automated scanning of filter surfaces for fluorescent objects | • Lack of sensitivity with chromosomal genes or mRNA as target<br>• Detection is strictly taxonomic<br>• Differentiation between living and dead cells is often difficult<br>• Not applicable to detect 1 indicator per 100 ml without concentration/filtration | • Potential for automation |

**Table 8.6. Methods for the detection of microbial contamination in drinking water (*continued*)**

| Method | Characteristics/advantages | Limitations/disadvantages | Application: status quo and future perspective |
|---|---|---|---|
| Molecular fingerprinting (ribotyping, RFLP, RAPD, AP-PCR) | • Faster than cultivation methods<br>• Excellent tool for differentiation of strains or isolates within a species | • At present no discrimination between viable and non-viable microorganisms<br>• RAPD requires the use of pure isolates | |
| DNA chip array | • Micromanufacturing techniques allows testing of up to several thousand sequences in one assay on a single "chip"<br>• Sensitive, selective and specific to the desired level to detect groups of organisms or (sub)-species, respectively<br>• Fast (2-4 hours) | • At present very cost intensive<br>• Highly trained personal needed<br>• Absolute quantitation might be problematic | • Technique not yet widely available |
| Biosensors | • Immunoaffinity step to bind microorganisms to surfaces; detection by laser excitation of bound fluorescent antibodies, acoustogravimetric wave transduction, or surface plasmon resonance<br>• Rapid, but depends on culturable microorganisms | | • Currently unable to discriminate between viable and non-viable microbes |
| Solid state biochip | • Aim of the method: rapid detection (minutes) of a number of toxins and microbial cells<br>• Approach does not require isolation and characterisation of nucleic acids | • Limitations cannot be determined yet | • Technique not yet available, visionary approach under development |

282

# REFERENCES

Abbaszadegan, M., Huber, M.S., Gerba, C.P. and Pepper, I.L. (1997) Detection of viable *Giardia* cysts by amplification of heat shock-induced mRNA. *Applied and Environmental Microbiology* **63**(1), 324-328.

Adams, M.H. (1959) Bacteriophages. Interscience, New York.

Amann, R.I., Krumholz, L. and Stahl, D.A. (1990) Fluorescent-oligonucleotide probing of whole cells for determinative, phylogenetic and environmental studies in microbiology. *Journal of Bacteriology* **172**(2), 762-70.

Amann, R.I., Ludwig, W. and Schleifer, K.-H. (1995) Phylogenetic identification and in situ detection of individual microbial cells without cultivation. *Microbiological Reviews* **59**, 143-169.

APHA, AWWA, WEF (1998) *Standard Methods for the Examination of Water and Wastewaters, 20ᵗʰ Edition.* American Public Health Association, Washington, DC.

Bej, A.K., Mahbubani, M.H. and Atlas, R.M. (1991) Detection of viable *Legionella pneumophila* in water by Polymerase Chain Reaction and gene probe methods. *Applied and Environmental Microbiology* **57**(2), 597-600.

Bej, A.K., Ng, W.Y., Morgan, S., Jones, D. and Mahbubani, M.H. (1996) Detection of viable *Vibrio cholerae* by reverse-transcription polymerase chain reaction (RT-PCR). *Molecular Biotechnology* **5**, 1-10.

Belasco, J. (1993) mRNA degradation in prokaryotic cells. In: *Control of Messenger RNA Stability.* Brawerman, B.A.G. (Ed.), Academic Press, Inc., San Diego, USA. pp. 3-12.

Calabrese, J.P. and Bissonnette, G.K. (1990) Improved membrane filtration method incorporating catalase and sodium pyruvate for detection of chlorine-stressed coliform bacteria. *Applied and Environmental Microbiology* **56**, 3558-3564.

Campbell, A.T., Robertson, L.J., Smith, H.V. and Girdwood, R.W.A. (1994) Viability of *Cryptosporidium parvum* oocysts concentrated by calcium carbonate flocculation. *Journal of Applied Bacteriology* **76**, 638-639.

CEN (2000) CEN/TC230/WG3/TG4. Enterovirus (Nr. 12), Working Document, Draft Method, March 2000.

Cochran, W.G. (1977) *Sampling Techniques, 3$^{rd}$ Edition.* John Wiley & Sons, New York.

Colwell, R.R. and Grimes, D.J. (Eds.) (2000) *Viable but Nonculturable Microorganisms in the Environment.* ASM Press, Washington, DC.

Deere, D., Vesey, G., Ashbolt, N. and Gauci, M. (2002) Flow cytometry and cell sorting for monitoring microbial cells. In: *Encyclopaedia of Environmental Microbiology.* Bitton, G. (Ed.). John Wiley and Sons, New York (in press).

Divizia, M., Santi, A.L. and Pana, A. (1989) Ultrafiltration: an efficient second step for hepatitis A virus and poliovirus concentration. *Journal of Virology Methods* **23**, 55-62.

Eckner, K.F. (1998) Comparison of membrane filtration and multiple-tube fermentation by the Colilert and Enterolert methods for detection of waterborne coliform bacteria, *Escherichia coli*, and enterococci used in drinking and bathing water quality monitoring in Southern Sweden. *Applied and Environmental Microbiology* **64**, 3079-3083.

Edberg, S.C., Allen, M.J. and Smith, D.B. (1991) Defined substrate technology method for rapid and specific simultaneous enumeration of total coliforms and *Escherichia coli* from water: collaborative study. *J. Assoc. Off. Analy. Chem.* **74**, 526-529.

Eggers, M.D., Balch, W.J., Mendoza, L.G., Gangadharan, R., Mallik, A.K., McMahon, M.G., Hogan, M.E., Xaio, D., Powdrill, T.R., Iverson, B., Fox, G.E., Willson, R.C., Maillard, K.I., Siefert, J.L. and Singh, N. (1997) Advanced approach to simultaneous monitoring of multiple bacteria in space. Chap. SAE Technical Series 972422. In: *27th International Conference on Environmental Systems, Lake Tahoe, Nevada, July 14-17,*

*1997*. The Engineering Society for Advancing Mobility Land Sea Air and Space, SAE International, Warrendale, PA, pp:1-8.

Ekins, R. and Chu, F.W. (1999) Microarrays: their origins and applications. *Tibtech* **17**, 217-218.

Emde, K.M.E., Mao, H. and Finch, G.R. (1992) Detection and occurrence of waterborne bacterial and viral pathogens. *Water and Environmental Research* **64**, 641-647.

Ericksen, T.H. and Dufour, A.P. (1986) Methods to identify waterborne pathogens and indicator organisms. In: *Waterborne Diseases in the United States*. G.F. Craun (Ed.). CRC Press, Boca Raton, USA. pp. 195-214.

Fricker, E.J., Illingworth, K.S. and Fricker, C.R. (1997) Use of two formulations of colilert and quantitray [TM] for assessment of the bacteriological quality of water. *Water Research* **31**, 2495-2499.

Fuchs, B.M., Wallner, G., Beisker, W., Schwippl, I., Ludwig, W. and Amann, R. (1998) Flow Cytometric Analysis of the in situ accessibility of *Escherichia coli* 16S rRNA for fluorescently labelled oligonucleotide probes. *Applied and Environmental Microbiology* **64**, 4973-4982.

Gasser, R.B. and Donaghue, P.O. (1999) Isolation, propagation and characterization of *Cryptosporidium. International Journal of Parasitology* **29**,1379-1413.

Georgia Tech Research Institute (1999) Georgia Tech Research News - September 1999.

Gonzalez, J.M. (1996) A general purpose program for obtaining most probable number tables. *Journal of Microbiological Methods* **26**(3), 215-218.

Goyal, S.M., Gerba, C.P. and Bitton, G. (Eds.) (1987) *Phage Ecology*. John Wiley and Sons, New York. pp 321.

Grabow, W.O.K., Holtzhausen, C.S. and De Villiers, J.C. (1993) *Research on Bacteriophages as Indicators of Water Quality.* WRC Report No 321/1/93. Water Research Commission, Pretoria. Pp. 147.

Grabow, W.O.K., Vrey, A., Uys, M. and De Villiers, J.C. (1998) *Evaluation of the Application of Bacteriophages as Indicators of Water Quality*. WRC Report No 540/1/98. Water Research. Commission, Pretoria.

Grimont, F. and Grimont, P.A.D. (1986) Ribosomal ribonucleic acid gene restriction patterns as potential taxonomic tools. *Annales de l'Institut Pasteur / Microbiology* **137B**, 165-175.

Grimont, P.A.D. (1988) Use of DNA reassociation in bacterial classification. *Canadian Journal of Microbiology* **34**, 541-546.

Grimont, P.A.D., Grimont, F., Desplaces, N. and Tchen, P. (1985) A DNA probe specific for *Legionella pneumophila*. *Journal of Clinical Microbiology* **21**, 431-437.

Guschin, D.Y., Mobarry, B.K., Proudnikov, D., Stahl, D.A., Rittmann, B.E. and Mirzabekov, A.D. (1997) Oligonucleotide microchips as genosensors for determinative and environmental studies in microbiology. *Applied and Environmental Microbiology* **63**, 2397-2402.

Horwitz, W. (1988) Protocol for the design, conduct and interpretation of collaborative studies. *Pure and Applied Chemistry* **60**, 855-864.

Hurst, C.J., Knudsen, G.R., McInerney, M.J., Stetzenbach, L.D. and Walter, M.V. (2001) *Manual of Environmental Microbiology, 2$^{nd}$ Edition*. American Society for Microbiology Press, Washington, DC.

ISO 10705-1 (1995) Water quality - Detection and enumeration of bacteriophages - Part 1: Enumeration of F-specific RNA bacteriophages. International Organization for Standardization, Geneva, Switzerland.

ISO 10705-2 (2000) Water quality - Detection and enumeration of bacteriophages - Part 2: Enumeration of somatic coliphages. International Organization for Standardization, Geneva, Switzerland.

ISO 10705-4 Water quality - Detection and enumeration of bacteriophages - Part 4: Enumeration of bacteriophages infecting *Bacteroides fragilis*. International Organization for Standardization, Geneva, Switzerland.

ISO 11731 (1998) Water quality - Detection and enumeration of *Legionella*. International Organization for Standardization, Geneva, Switzerland.

ISO 6222 (1999) Water quality - Enumeration of culturable microorganisms - Colony count by inoculation in a nutrient agar culture medium. International Organization for Standardization, Geneva, Switzerland.

ISO 6340 (1995) Water quality - Detection and enumeration of *Salmonella*. International Organization for Standardization, Geneva, Switzerland.

ISO 6461-1 (1986) Water quality - Detection and enumeration of the spores of sulfite-reducing anaerobes (clostridia) - Part 1: Method by enrichment in a liquid medium. International Organization for Standardization, Geneva, Switzerland.

ISO 6461-2 (1986) Water quality - Detection and enumeration of the spores of sulfite-reducing anaerobes (clostridia) - Part 2: Method by membrane filtration. International Organization for Standardization, Geneva, Switzerland.

ISO 7704 (1985) Water quality - Evaluation of membrane filters used for microbiological analyses. International Organization for Standardization, Geneva, Switzerland.

ISO 7899-1 (1998) Water quality - Detection and enumeration of intestinal enterococci in surface and waste water - Part 1: Miniaturized method (Most Probable Number) by inoculation in liquid medium. International Organization for Standardization, Geneva, Switzerland.

ISO 7899-2 (2000) Water quality - Detection and enumeration of intestinal enterococci - Part 2: Membrane filtration method. International Organization for Standardization, Geneva, Switzerland.

ISO 8360-1 (1998) Water quality - Detection and enumeration of *Pseudomonas aeruginosa* - Part 1: Method by enrichment in liquid medium. International Organization for Standardization, Geneva, Switzerland.

ISO 8360-2 (1998) Water quality - Detection and enumeration of *Pseudomonas aeruginosa* - Part 2: Membrane filtration method. International Organization for Standardization, Geneva, Switzerland.

ISO 9308-1 (2000) Water quality - Detection and enumeration of *Escherichia coli* and coliform bacteria - Part 1: Membrane filtration method. International Organization for Standardization, Geneva, Switzerland.

ISO 9308-2 (1990) Water quality - Detection and enumeration of coliform organisms, thermotolerant coliform organisms and presumptive *Escherichia coli* - Part 2: Multiple tube (most probable number) method. International Organization for Standardization, Geneva, Switzerland.

ISO 9998 (1991) Water quality - Practices for evaluating and controlling microbiological colony count media used in water quality tests. International Organization for Standardization, Geneva, Switzerland.

ISO/CD 17994 Water quality - Criteria for the establishment of equivalence between microbiological methods. International Organization for Standardization, Geneva, Switzerland.

ISO/DIS 10705-3 Water quality - Detection and enumeration of bacteriophages - Part 3: Validation of methods for concentration of bacteriophages from water. International Organization for Standardization, Geneva, Switzerland.

ISO/DIS 11731-2 Water quality - Detection and enumeration of *Legionella* - Part 2: Direct membrane filtration method for waters with low bacterial counts. International Organization for Standardization, Geneva, Switzerland.

ISO/TR 13843 (2000) Water quality - Guidance on validation of microbiological methods. International Organization for Standardization, Geneva, Switzerland.

Kalmbach, S., Manz, W. and Szewzyk, U. (1997) Dynamics of biofilm formation in drinking water: phylogenetic affiliation and metabolic potential of single cells assessed by formazan reduction and in situ hybridization. *FEMS Microbial Ecology* **22**, 265-279.

Kenedy, Jr., J.J., Bitton, G. and Oblinger, J.L. (1985) Comparison of selective media for assay for coliphages in sewage effluent and lake water. *Applied and Environmental Microbiology* **49**, 33-36.

Kilger, G. and Grimont, P.A.D. (1993) Differentiation of *Salmonella* Phase 1 flagellar antigen types by restriction of the amplified *fliC* gene. *Journal of Clinical of Microbiology* **31**, 1108-1110.

Kogure, K., Simidu, U. and Targa, N. (1979) A tentative direct microscopic method for counting living marine bacteria. *Canadian Journal of Microbiology* **25**, 415-420.

LeChevallier, M.W., Cameron, S.C. and McFeters, G.A. (1982) New medium for improved recovery of coliform bacteria from drinking water. *Applied and Environmental Microbiology* **45**, 484-492.

Lee, R.M. and Hartman, P.A. (1989) Optimal pyruvate concentration for the recovery of coliforms from food and water. *Journal of Food Protection* **52**, 119-121.

Lemieux, B., Aharoni, A. and Schena, M. (1998) Overview of DNA chip technology. *Mol. Breeding* **4**, 277-289.

Lightfoot, N.F. and Maier, E.A. (Eds.) (1988) *Microbiological Analysis of Food and Water. Guidelines for Quality Assurance.* Elsevier, Amsterdam.

Lipshutz, R.J., Fodor, S.P.A., Gingeras, T.R. and Lockhart, D.J. (1999) High density synthetic oligonucleotide arrays Review. *Nature Genetics* **21**(Suppl S), 20-24.

Manafi, M. (1996) Fluorogenic and chromogenic substrates in culture media and identification tests. *International Journal of Food Microbiology* **31**, 45-58.

McClure, F.D. (1990) Design and analysis of qualitative collaborative studies: minimum collaborative program. *J. Assoc. Off. Anal. Chem.* **73**, 953-960.

McFeters, G.A., Kippin, J.S. and LeChevallier, M.W. (1986) Injured coliforms in drinking water. *Applied and Environmental Microbiology* **51**, 1-5.

Nishimura, M., Kita-Tsukamoto, K., Kogure, K., Ohwada, K. and Simidu, U. (1993) A new method to detect viable bacteria in natural seawater using 16S rRNA oligonucleotide probe. *Journal of Oceanology* **49**, 51-57.

Ongerth, J.E. and Stibbs, H.H. (1987) Identification of *Cryptosporidium* oocysts in river water. *Applied and Environmental Microbiology* **53**, 672-676.

Padan, E., Shilo, M. and Kislev, N. (1967) Isolation of "cyanophage" from freshwater ponds and their interaction with *Plectonema boryanum*. *Virology* **32**, 234-246.

Payment, P. (2001) Cultivation of viruses from environmental samples. In: *Manual of Environmental Microbiology, 2nd Edition.* Hurst, C.J., Knudsen, G.R., McInerney, M.J., Stetzenbach, L.D. and Walter, M.V. (Eds.) American Society for Microbiology Press, Washington, DC.

Prescott, A.M. and Fricker, C.R. (1999) Use of PNA oligonucleotides for the in situ detection of *Escherichia coli* in water. *Molecular and Cellular Probes* **13**, 261-268.

Regnault, B., Martin-Delautre, S., Lejay-Collin, M., Lefèvre, M. and Grimont, P.A.D. (2000) Oligonucleotide probe for the visualization of *Escherichia coli/E. fergusonii* cells by in situ hybridization: specificity and potential applications. *Research in Microbiology* **151**, 521-533.

Rice, E.W., Allen, M.J., Covert, T.C., Langewis, J. and Standridge, J. (1993) Identifying *Escherichia* species with biochemical test kits and standard bacteriological tests. *Journal of the American Water Works Association* **85**, 74-76.

Rijpens, N.P., Jannes, G., van Asbroek, M., Herman, L.M.F. and Rossau, R. (1995) Simultaneous detection of *Listeria* spp. and *Listeria monocytogenes* by reverse hybridization with 16S-23S rRNA spacer probes. *Molecular and Cellular Probes* **9**, 423-432.

Roszak, D.B. and Colwell, R.R. (1987) Metabolic activity of bacterial cells enumerated by direct viable count. *Applied and Environmental Microbiology* **53**, 2889-2893.

Schiemann, D.A. (1990) *Yersinia enterocolitica* in drinking water. In: *Drinking Water Microbiology: Progress and Recent Developments.* McFeters, G.A. (Ed.), Springer-Verlag, New York. , pp. 428-451.

Seeley, N.D. and Primrose, S.B. (1982) The isolation of bacteriophages from the environment. *Journal of Applied Bacteriology* **53**, 1-17.

Sheridan, G.E.C., Masters, C.I., Shallcross, J.A. and Mackey, B.M. (1998) Detection of mRNA by Reverse Transcription-PCR as an indicator of viability in *Escherichia coli* cells. *Applied and Environmental Microbiology* **64**(4), 1313-1318.

Shieh, Y.S.C., Baric, R.S. and Sobsey, M.D. (1997) Detection of low levels of enteric viruses in metropolitan and airplane sewage. *Applied and Environmental Microbiology* **62**, 4401-4407.

Shields, P.A. and Farrah, S.R. (1986) Concentration of viruses in beef extract by flocculation with ammonium sulfate. *Applied and Environmental Microbiology* **51**, 211-213.

Simmons, O.D., Sobsey, M.D., Heaney, C.D., Schaefer III, F.W. and Francy, D.S. (2001) Concentration and detection of *Cryptosporidium* oocysts in surface water samples by method 1622 using ultrafiltration and capsule filtration. *Applied and Environmental Microbiology* **67**, 1123-1127.

Singh, S.N. and Gerba, C.P. (1983) Concentration of coliphage from water and sewage with charge-modified filter aid. *Applied and Environmental Microbiology* **45**, 232-237.

Slifko, T.R., Friedman, D., Rose, J.B. and Jakubowski, W. (1997) An *in vitro* method for detecting infectious *Cryptosporidium* oocysts with cell culture. *Applied and Environmental Microbiology* **63**(9), 3669-75.

Sobsey, M.D. and Glass, J.S. (1980) Poliovirus concentration from tapwater with electro-positive adsorbent filters. *Applied and Environmental Microbiology* **40**, 201-210.

Sobsey, M.D., Schwab, K.J. and Handzel, T.R. (1990) A simple membrane filter method to concentrate and enumerate male-specific RNA coliphages. *Journal of the American Water Works Association* **82**, 52-59.

States, S.J., Wadowsky, R.M., Kuchta, J.M., Wolford, R.S., Conley, L.F. and Yee, R.B. (1990) Legionella in drinking water. In: *Drinking Water Microbiology: Progress and Recent Developments.* McFeters, G.A. (Ed.), Springer-Verlag, New York. pp. 340-367.

Stinear, T., Matusan, A., Hines, K. and Sandery, M. (1996) Detection of a single viable *Cryptosporidium parvum* oocyst in environmental water concentrate by reverse transcription-PCR. *Applied and Environmental Microbiology* **62**, 3385-3390.

Tillett, H.E. and Lightfoot, N.F. (1995) Quality control in environmental microbiology compared with chemistry: What is homogeneous and what is random? *Water Science and Technology* **31**, 471-477.

Trebesius, K., Amann, R., Ludwig, W., Mühlegger, K. and Scheleifer, K.-H. (1994) Identification of whole fixed bacterial cells with nonradioactive 23 S rRNA-targeted polynucleotide probes. *Applied and Environmental Microbiology* **60**, 3228-3235.

USEPA (1999) Method 1623 - *Cryptosporidium* and *Giardia* in Water by Filtration/IMS/IFA. (EPA-821-R99-006.) Office of Water, United States

Environment Protection Agency, Washington D.C.
http://www.epa.gov/nerlcwww/1623.pdf pages.

Vesey, G., Ashbolt, N., Fricker, E.J., Deere, D., Williams, K.L., Veal, D.A. and
Dorson, M. (1998) The use of ribosomal RNA targeted oligonucleotide
probe for fluorescent labelling of viable *Cryptosporidium parvum*
oocysts. *Journal of Applied Microbiology* **85**, 429-440.

Vesey, G., Narai, J., Ashbolt, N., Williams, K. L. and Veal, D. A. (1994)
Detection of specific microorganisms in environmental samples using
flow cytometry. In: *Methods in Cell Biology*. Darzynkiewicz, Z.,
Robinson, J. P. and Crissman, H. A. (Eds.). Vol. 42. Academic Press Inc.,
New York, pp. 489-522.

Vesey, G., Slade, J.S., Byrne, M., Shepherd, K. and Fricker, C.R. (1993) A new
method for the concentration of *Cryptosporidium* oocysts from water.
*Journal of Applied Bacteriology* **75**(1), 82-86.

Villarino, A., Bouvet, O.M.M., Regnault, B., Martin Delautre, S. and Grimont,
P.A.D. (2000) Exploring the frontier between life and death in
Escherichia coli: evaluation of different viability markers in live and
heat- and UV-killed cells. *Research in Microbiology* **151**, 755-768.

Wang, J., Rivas, G., Cai, X., Palecek, E., Nielsen, P., Shiraishi, H., Dontha, N.,
Luo, D., Parrado, C., Chicharro, M., Farias, P.A.M., Valera, F.S., Grant,
D.H., Ozsoz, M. and Flair, M.N. (1997) DNA electrochemical biosensors
for environmental monitoring – a review. *Analt. Chimica Acta* **347**, 1-8.

Watkins, J. and Xiangrong, J. (1997) Cultural methods of detection for
microorganisms: recent advances and successes. In: *The Microbiological
Quality of Water*. Sutcliffe, D.W. (Ed.) Freshwater Biological Association,
Ambleside. pp:19-27.

WHO (1984) *Guidelines for Drinking water Quality, Vol. 2. Health criteria and
other supporting information*. World Health Organization, Geneva.

Wilson, I.G. (1997) Inhibition and facilitation of nucleic acid amplification.
Mini review. *Applied and Environmental Microbiology* **63**, 3741-3751.

Xu, H.-S., Roberts, N., Singleton, F.L., Attwell, R.W., Grimes, D.J., Colwell,
R.R. (1982) Survival and viability of nonculturable *Escherichia coli* and
*Vibrio cholerae* in the estuarine and marine environment. *Microbial
Ecology* **8**, 313-323.

# LIST OF ABBREVIATIONS

| | |
|---|---|
| AFLP | Amplified fragment length polymorphism |
| APHA | American Public Health Association |
| AP-PCR | Arbitrarily primed polymerase chain reaction |
| AQC | Analytical quality control |
| ATCC | American Type Culture Collection, Manassas, VA, USA |
| BCYE | Buffered charcoal yeast extract agar medium |
| BCYE-cys | BCYE without L-cystein |
| BEAA | Bile esculin azide agar |
| BGBB | Brilliant green bile broth |
| BGM | Buffalo green monkey |
| BPRM | Bacteroides phage recovery medium |
| cDNA | Complementary DNA |
| CEN | European Committee for Standardisation |
| cfu | Colony forming unit |
| DAL | Double agar layer |
| DAS-ELISA | Double antibody sandwich enzyme linked immunosorbent assay |
| DNA | Deoxyribonucleic acid |
| DRCM | Clostridium differential bouillon |

| | |
|---|---|
| DVC | Direct viable count |
| EC | Selective broth for the growth of coliforms and *E. coli* , containing bile salts |
| EIA | Enzyme immunoassay |
| ELISA | Enzyme linked immunosorbent assay |
| FACS | Fluorescently activated cell sorting |
| FISH | Fluorescent in-situ hybridisation |
| GIT | Guanidinium isothiocyanate |
| GVPC | BCYE with *Legionella*-selective supplements |
| HAV | Hepatitis A virus |
| HRP | Horseradish peroxidase |
| ICC RT-PCR | Integrated cell culture reverse transcriptase polymerase chain reaction |
| IMS | Immunomagnetic separation |
| ISO | International Organization for Standardization |
| MPN | Most probable number |
| mRNA | Messenger RNA |
| MSA | Modified Scholtens' agar |
| MSB | Modified Scholtens' broth |
| MUG | 5-methylumbelliferyl-$\beta$-D-glucuronide |
| NCTC | National Collection of Type Cultures and Pathogenic Fungi, (Public Health Laboratory Service, UK) |
| ONPG | O-nitrophenyl-$\beta$-D-galactopyranoside |
| PCR | Polymerase chain reaction |
| PEG | Polyethylene glycol |

| | |
|---|---|
| pfu | Plaque forming unit |
| RAPD | Random amplified polymorphic DNA |
| RNA | Ribonucleic acid |
| rRNA | Ribosomal RNA |
| RT-PCR | Reverse transcriptase polymerase chain reaction |
| $TCID_{50}$ | Tissue culture for 50% infective dose |
| tRNA | Transfer RNA |
| TSA | Tyramide signal amplification |
| TYGA | Tryptone yeast extract-glucose agar |
| TYGB | Tryptone yeast extract-glucose broth |
| UV | Ultra violet |
| VBNC | Viable but nonculturable |

OECD PUBLICATIONS, 2, rue André-Pascal, 75775 PARIS CEDEX 16
PRINTED IN FRANCE
(93 2003 01 1 P) ISBN 92-64-09946-8 – No. 52783 2003